A HANDS-ON INTRODUCTION TO

SOLIDWORKS 2023

Kirstie Plantenberg
University of Detroit Mercy

SDC
PUBLICATIONS

SDC Publications
P.O. Box 1334
Mission, KS 66222
913-262-2664
www.SDCpublications.com
Publisher: Stephen Schroff

Trademarks
SOLIDWORKS®, eDrawings®, SOLIDWORKS Simulation®, SOLIDWORKS Flow Simulation, and SOLIDWORKS Sustainability are registered trademarks of Dassault Systèmes SOLIDWORKS Corporation in the United States and other countries; certain images of the models in this publication courtesy of Dassault Systèmes SOLIDWORKS Corporation.

The author and publisher of this book have used their best efforts in preparing this book. These efforts include the development, research and testing of the material presented. The author and publisher shall not be liable in any event for incidental or consequential damages with, or arising out of, the furnishing, performance, or use of the material.

ISBN-13: 978-1-63057-555-7
ISBN-10: 1-63057-555-0

Printed and bound in the United States of America.

A HANDS-ON INTRODUCTION TO SOLIDWORKS®

PREFACE

Overview

A Hands-On Introduction to SOLIDWORKS® is specifically written for those who are new to SOLIDWORKS®. This book allows you to learn as you work through tutorials that show you the basics of the software and work your way up to more in-depth skills. Formerly called *Project Based SOLIDWORKS®*, this revised edition includes new and expanded tutorials. *A Hands-On Introduction to SOLIDWORKS®* works perfectly for a freshman design class or as a companion text to an engineering graphics textbook.

Many tutorials in the book connect 3-D modeling and engineering graphics concepts. Learn how to model parts, configurations, create part prints, assemblies, and assembly drawings. As you become more comfortable with SOLIDWORKS®, later chapters introduce FEA, how to create more complex solid geometries with parametric modeling, apply tolerances, and use advanced and mechanical mates. Important commands and features are highlighted and defined in each chapter to help you become familiar with them.

Student Resources

Many students prefer learning software through video instruction. Therefore, this book has **instructional videos** showing how to perform each tutorial. It also comes with instructional videos showing how to complete each end-of-chapter problem. Some problems are purposely left open-ended to simulate real-life design situations; therefore, more than one solution is possible. After completing all the tutorials in this book, you will be able to accurately design moderately difficult parts and assemblies and have a firm foundation in SOLIDWORKS®.

A Hands-On Introduction to SOLIDWORKS® comes with all the part/assembly files that a student will need to complete the tutorials.

Topics covered

The following topics are covered in this book.

- Part modeling
- Part configurations
- Assembly
- Static FEA
- Part Prints
- Assembly drawings
- Fasteners
- Tolerancing
- Parametric Modeling
- 3D-sketches
- Rendering

Instructor Supplements

      A Hands-On Introduction to SOLIDWORKS® comes with a set of quiz problems not included in the book itself. Quiz files will be provided to instructors. These are the files that students will need in order to complete some of the quizzes.

Have questions? E-mail: plantenk@udmercy.edu
                    Please include the book and edition that the question refers to.

Found a mistake? Please e-mail a detailed description of the errata to:
                    plantenk@udmercy.edu
                    Please include the book and edition that the errata refers to.

This book is dedicated to my family for their support and help.

I would also like to thank

Alessandra Pontoni and Alena Krolczyk for the use of their awesome model shown on the cover of this book.

Joseph Jessop, Paul Dellock, and Job Gumma for the use of their models as quiz problems in the book.

A HANDS-ON INTRODUCTION TO SOLIDWORKS®

TABLE OF CONTENTS

CHAPTER 1: BASIC MODELING IN SOLIDWORKS®

- Cylinder Tutorial
- Angled Block Tutorial
- Connecting Rod Tutorial

CHAPTER 2: BASIC PART PRINTS IN SOLIDWORKS®

- Angled Block Part Print Tutorial
- Defining a Front View Tutorial

CHAPTER 3: INTERMEDIATE PART MODELING IN SOLIDWORKS®

- Flanged Coupling Tutorial
- Tabletop Mirror Tutorial

CHAPTER 4: INTERMEDIATE PART PRINTS IN SOLIDWORKS®

- Connecting Rod Part Print Tutorial
- Flanged Coupling Part Print Tutorial

CHAPTER 5: CONFIGURATIONS IN SOLIDWORKS®

- Connecting Rod Configurations Tutorial
- Flanged Coupling Configurations Tutorial
- Rigid Coupling Design Table Tutorial

CHAPTER 6: FEA IN SOLIDWORKS®

- Connecting Rod FEA Tutorial

CHAPTER 7: BASIC ASSEMBLIES IN SOLIDWORKS®

- Flanged Coupling Assembly Tutorial

CHAPTER 8: ASSEMBLY DRAWINGS IN SOLIDWORKS®

- Flanged Coupling Assembly Drawing Tutorial

CHAPTER 9: ADVANCED MODELING IN SOLIDWORKS®

- Microphone Base Tutorial
- Microphone Arm Tutorial
- Boat Tutorial

CHAPTER 10: INTERMEDIATE ASSEMBLY IN SOLIDWORKS®

- Tabletop Mirror Assembly Tutorial
- Microphone Assembly Tutorial
- Linear Bearing Assembly Tutorial

CHAPTER 11: TOLERANCING AND THREADS IN SOLIDWORKS®

- Vise – Stationary Jaw Tutorial
- Vise – Screw Tutorial

CHAPTER 12: PARAMETRIC MODELING IN SOLIDWORKS®

- Vise – Spacer Tutorial
- Vise – Jaw Insert Tutorial

CHAPTER 13: ADVANCED ASSEMBLY IN SOLIDWORKS®

- Vise Assembly Tutorial
- Gear Assembly Tutorial

CHAPTER 14: 3D SKETCHES IN SOLIDWORKS®

- Handlebar Tutorial

CHAPTER 15: RENDERING IN SOLIDWORKS®

- Glass Vase Tutorial
- Light Bulb Tutorial

CHAPTER 1

BASIC PART MODELING IN SOLIDWORKS®

CHAPTER OUTLINE

1.1) WHAT IS SOLIDWORKS® 3
1.2) STARTING SOLIDWORKS, USER INTERFACE, AND FILE TYPE...........................4
    1.2.1) Starting **SOLIDWORKS®** 4
    1.2.2) **SOLIDWORKS®** user interface...6
    1.2.3) File types...8
1.3) NEW PART, USER INTERFACE, AND SET UP 9
    1.3.1) New Part 9
    1.3.2) Part user interface..10
    1.3.3) Setting up your part file 11
    1.3.4) Saving your part file 13
1.4) SKETCHING..13
    1.4.1) Sketching on a plane or face 14
    1.4.2) Editing a sketching...14
    1.4.3) Lines 15
    1.4.4) Rectangles 15
    1.4.5) Circles 16
    1.4.6) Arcs 16
    1.4.7) Slots 16
    1.4.8) Polygon 17
    1.4.9) Sketch Chamfer/Fillet 18
1.5) DIMENSIONS AND RELATIONS 19
    1.5.1) Sketch relations 19
    1.5.2) Smart Dimension 20
    1.5.3) Editing a Dimension 20
1.6) FEATURES...22
    1.6.1) Extrude and Extrude cut 22
    1.6.2) Fillet and Chamfer...24
1.7) FEATURE MANAGER DESIGN TREE..26
    1.7.1) Editing a Feature..27
    1.7.2) Editing a Sketch 27
    1.7.3) Rollback Bar...27
1.8) VIEWING YOUR PART 28
1.9) ASSIGNING MATERIAL 30
1.10) EVALUATE..31

1.11) CYLINDER TUTORIAL **33**

1.11.1) Prerequisites 33

1.11.2) What you will learn? 33

1.11.3) Setting up the part.. 34

1.11.4) Sketching and Extrude Boss/Base 39

1.11.5) Sketching and extrude cut 42

1.12) ANGLED BLOCK TUTORIAL.. **44**

1.12.1) Prerequisites 44

1.12.2) What you will learn? 44

1.12.3) Setting up the project 45

1.12.4) Creating the part 45

1.12.5) Adding material 49

1.13) CONNECTING ROD TUTORIAL **51**

1.13.1) Prerequisites 51

1.13.2) What you will learn 51

1.13.3) Setting up the project 52

1.13.4) Base extrude 53

1.13.5) Adding features 57

1.13.6) Adding material 59

BASIC MODELING IN SOLIDWORKS® PROBLEMS **61**

CHAPTER SUMMARY

In this chapter, you will learn how to navigate through and create basic models in SOLIDWORKS®. Sketch elements such as Lines, Circles, Rectangles, etc. will be explored. Features such as Extrude can then be used to create solids using these sketch elements. Although SOLIDWORKS® may appear complex, by the end of this book you will feel very comfortable navigating within its different workspaces. By the end of this chapter, you will be able to create and edit simple sketch elements, dimensions, and features.

1.1) WHAT IS SOLIDWORKS®

SOLIDWORKS® is a solid modeling computer-aided design and computer-aided engineering program. Computer-aided design (CAD) uses computer software to create, modify, analyze, and optimize a design. The term CADD (Computer-aided design and drafting) may also be used. The advantage of using CAD over the traditional design-prototype-test cycle of design is speed and lower cost. Computer-aided design and computer-aided engineering are sometimes not used to mean the same thing. Computer-aided engineering (CAE) uses CAD software to analyze the components and assemblies through simulation, optimization, manufacturing along with other tools.

This text will focus mainly on the modeling of parts and assemblies along with creating technically correct 2D drawings of these components. Figure 1.1-1 shows examples of a part and its associated drawing created in SOLIDWORKS®.

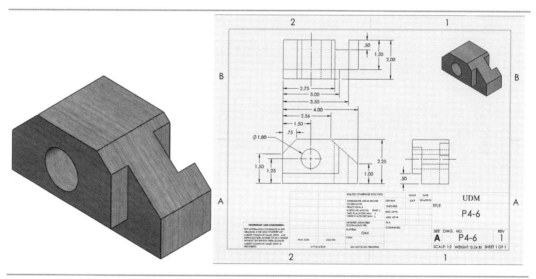

Figure 1.1-1: SOLIDWORKS examples

1.2) STARTING SOLIDWORKS, USER INTERFACE, AND FILE TYPE

1.2.1) Starting SOLIDWORKS®

When you double-click on the SOLIDWORKS icon ![SW 2021], a *Welcome – SOLIDWORKS* window will appear. Notice that there are several tabs.

- **Home:** The *Home* tab, shown in Figure 1.2-1, is sort of a catch all. This is where you can access recently worked on models, start a new part, assembly or drawing, or access resources.
- **Recent:** The *Recent* tab shows your most recently worked on parts, assemblies and drawings.
- **Learn:** The *Learn* tab, shown in Figure 1.2.-2, is where you can access SOLIDWORKS information, tutorials and training.
- **Alerts:** The Alerts tab, shown in Figure 1.2-3, is where you can access software updates.

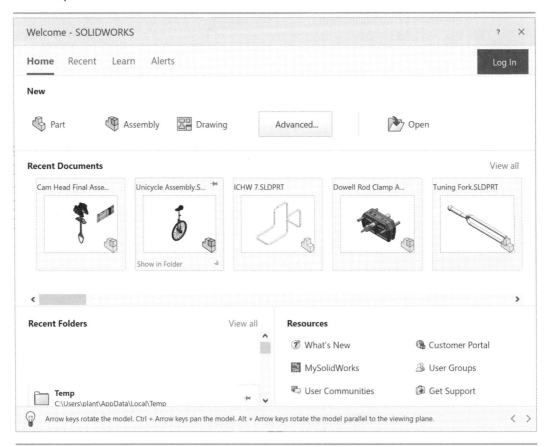

Figure 1.2-1: Welcome to SOLIDWORKS® - Home Tab

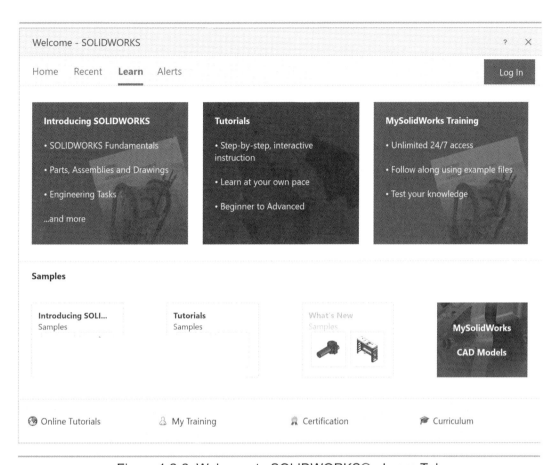

Figure 1.2-2: Welcome to SOLIDWORKS® - Learn Tab

Figure 1.2-3: Welcome to SOLIDWORKS® - Alerts Tab

1.2.2) SOLIDWORKS® user interface

If you close the Welcome to SOLIDWORKS® window without starting a new part, assembly or drawing, the SOLIDWORKS® user interface should look similar to what is shown in Figure 1.2.-4. There are a few areas where you should focus your concentration.

- **Quick Access Toolbar:** This is where you can access commands that are used frequently.
- **Pull-down Menu:** The pull-down menus may be accessed by clicking on the black triangle just to the right of the SOLIDWORKS® logo in the top left corner (See Figure 1.2-5). You can pin the pull-down menu by clicking on the pushpin icon. Many of the commands that are usually accessed by clicking on icons in the ribbon, can be accessed through the pull-down menu. In fact, some commands can only be accessed through the pull-down menu.
- **Help:** If you are having trouble performing a particular task in SOLIDWORKS®, click on the help icon ⌐⑦⌐ and you will be brought to the online help window shown in Figure 1.2-6. This site allows you to search specific commands and retrieve information explaining the use of each command. You can also type a keyword in the search box.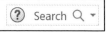
- **Task Pane:** This is where you can access the following (from top to bottom).
  - 3D Experience Marketplace
  - SOLIDWORKS® Resources such as tutorials.
  - Design Library where standard parts may be accessed.
  - File Explorer
  - View Palette
  - Appearance, Scenes and Decals where you can change the look of your part.
  - Custom Properties

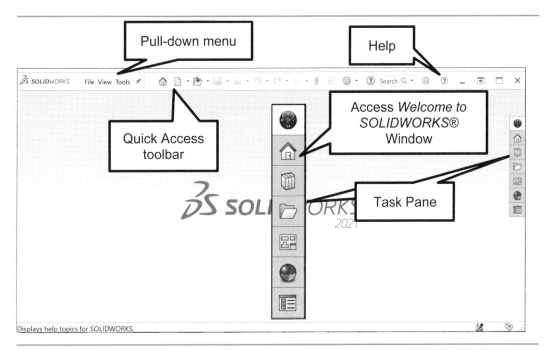

Figure 1.2-4: SOLIDWORKS® user interface

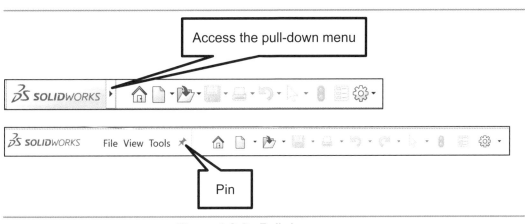

Figure 1.2-5: Pull-down menu

Figure 1.2-6: Online SOLIDWORKS® help

1.2.3) File types

SOLIDWORKS® has three main file types: Part file, Assembly file, and Drawing file.

- **Part file:** A part file consists of one single part. The file extension is *PartName.SLDPRT*. A person with the same version of SOLIDWORKS® or a later version of SOLIDWORKS® can open your file. However, a person with an older version of SOLIDWORKS will not be able to open your file. There is not backwards compatibility or a way to save a SOLIDWORKS® part file as an older version.
- **Assembly file:** An assembly file consists of several parts put together to make an assembly. Think of a bicycle which consists of many parts. The file extension for an assembly is *AssemblyName.SLDASM*. In order for someone to open your assembly, they will need the assembly file plus all the part files that make up the assembly.
- **Drawing file:** A drawing file is a 2-D drawing of a part or an assembly. The file extension for a drawing is *DrawingName.SLDDRW*. In order for someone to open and view your drawing, they will need every part and assembly file that was used to create the drawing plus the drawing file.

SOLIDWORKS® allows you to export to many different file types including 3dxml, step, stl, eprt, cgr, pdf, jpg and many more. In a pinch, this can be a way to share files if you and your partner don't have the same version of SOLIDWORKS®.

1.3) NEW PART, USER INTERFACE, AND SET UP

1.3.1) New Part

You start a new part when you want to model a single object. There are a couple ways you can begin a new part. If you are just starting SOLIDWORKS®, you can select the part icon in the *Welcome to SOLIDWORKS* window (see Figure 1.3-1). If you are already in SOLIDWORKS®, you can use the following methods to start a new part.

- Click on the **New Part** icon ⬜ from the *Quick Access* toolbar at the top.
- Short cut key: **Ctrl + N**
- Pull-down menu: **File – New…**

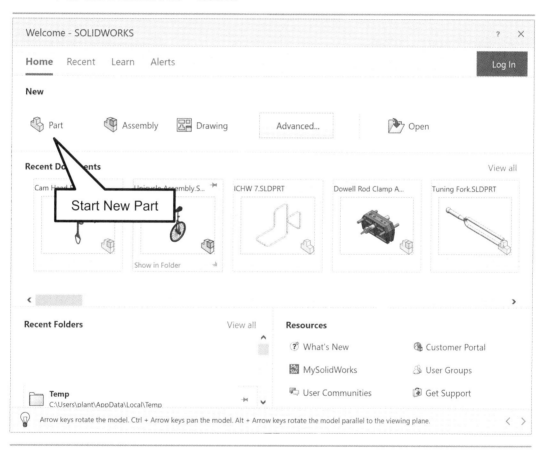

Figure 1.3-1: Welcome to SOLIDWORKS® - Home Tab – New Part

1.3.2) Part user interface

When you start a new part, you will enter the part user interface. This will have some differences from the user interface used when you create an assembly or drawing. The main areas that you should get familiar with are:

- **Command Manager:** The *Command Manager* contains most of the commands needed to construct your model. This area will change depending on what type of file you are editing (e.g. part, assembly or drawing).
- **Feature Manager Design Tree:** This is where all the operations/features, planes and other reference geometry are listed in the order of creation.
- **View (heads-up) toolbar:** The commands located in this toolbar allow you to manipulate how you view your model.
- **Status bar:** The status bar will display text that gives you instructions on how to complete a command.
- **Quick unit switching:** This area allows you to quickly change between systems of units.

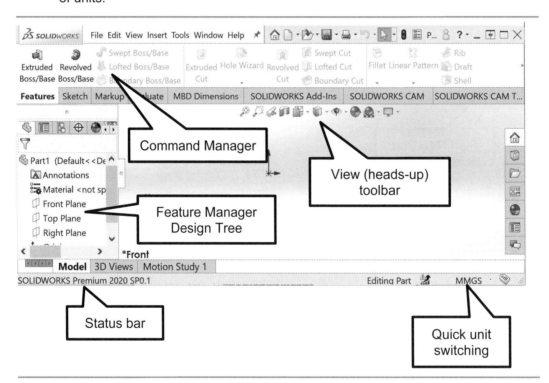

Figure 1.3-2: Part user interface

1.3.3) Setting up your part file

Before you start modeling your part, you should set the drafting standard and units.

To set the standard and units select the **Options** 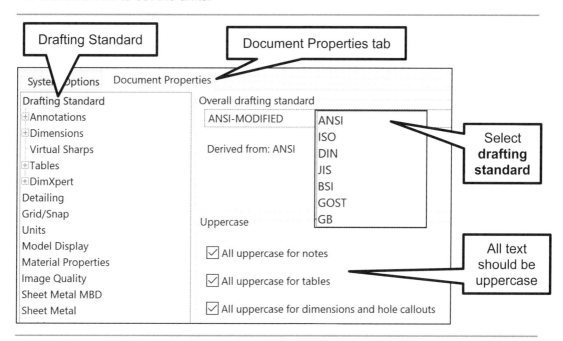 command located at the top of the drawing area.   Figure 1.3-3 shows how to set the drafting standard and Figure 1.3-4 and 1.3-5 shows how to set the units.

Figure 1.3-3: Setting the Drafting Standard

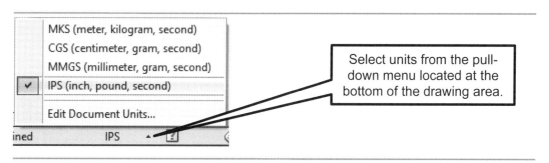

Figure 1.3-4: Setting Units

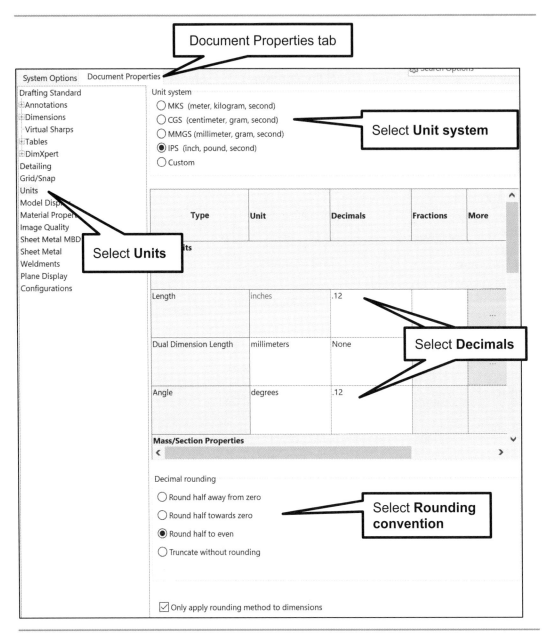

Figure 1.3-5: Setting Units

1.3.4) Saving your part file

You should also save your file initially and save often throughout the modeling process. You can never predict when your computer or the program will shut down. SOLIDWORKS® uses different file extensions depending on what type of file you are saving.

- Part = *.SLDPRT
- Assembly = *.SLDASM
- Drawing = *.SLDDRW

1.4) SKETCHING

In general, before you can create a solid, you need to create a two-dimensional sketch that will define the solid's shape. In order to sketch the two-dimensional element, you need to select a plane or face to start the sketch. You can do this in a few different ways. When you start sketching, it is very helpful to look at the sketch plane straight on. You can view the sketch plane straight on by using the short cut key **Ctrl+8**. Many different sketch elements are available in the *Sketch* tab shown in Figure 1.4-1. Once you have created the sketch, it will show up in your *Feature Design Tree*. From there you can edit the sketch if later you find that you need to modify it. See the informational box to learn how to select a sketch plane or face and edit a sketch.

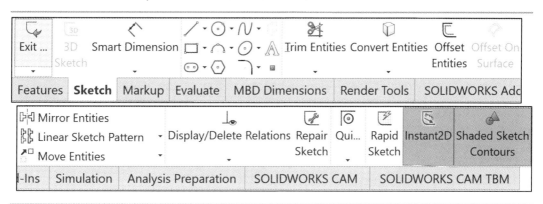

Figure 1.4-1: Sketch tab commands

1.4.1) Sketching on a plane or face

There are generally three different ways to start a sketch on a plane or a face. Use the following steps to start a sketch.

Method 1

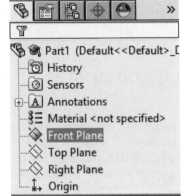

1) Select a plane in the *Feature Manager Design Tree* or a face on your part.

2) Select **Sketch**  from the *Command Manager - Sketch* tab.

Method 2

1) Select a plane in the *Feature Manager Design Tree* or a face on your part. A *Context* toolbar should appear.

2) Select **Sketch** ⌐ from the *Context* toolbar.

Method 3

1) Make sure everything is deselected by hitting the **Esc** button a couple times.

2) Select **Sketch** from the *Command Manager - Sketch* tab.
3) Select the plane or face where you wish to sketch.

1.4.2) Editing a sketching

You can edit a sketch even after a feature has been applied to it. Use the following steps to edit a sketch.

1) Select the sketch to be edited in the *Feature Manager Design Tree*.

2) Select **Edit Sketch** from the *Context* toolbar.

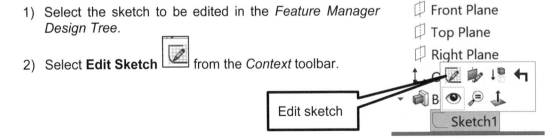

1.4.3) Lines

You are able to draw different line types. The line commands are located in the *Sketch* **tab** in the *Command Manager*. A description of each line type is given below.

- **Line:** [/ Line] A *Line* is created by selecting its start point and then its end point.
    - ○ <u>Single line:</u> Click (start point) – Hold – Drag – Release (end point)
    - ○ <u>Connected lines:</u> Click (Start point) – Release – Click (end point)
- **Centerline:** [⟋ Centerline] A *Centerline* is created exactly like a line, but it is a *Construction Geometry*. This means that it only helps you draw and is not a line that defines an edge of your solid part. However, it is often used to define axes of rotation for commands such as *Revolve*.
- **Midpoint Line:** [⟍ Midpoint Line] A *Midpoint Line* is used to create a line by selecting the midpoint of the line first and then selecting the orientation of the line.

1.4.4) Rectangles

You are able to draw different rectangle types. The rectangle commands are located in the *Sketch* **tab** in the *Command Manager*. A description of each rectangle type is given below.

- **Corner Rectangle:** [▢ Corner Rectangle] A *Corner Rectangle* is defined by two corner points.
- **Center Rectangle:** [▣ Center Rectangle] A *Center Rectangle* is defined by a geometric center and a corner point.
- **3 Point Corner Rectangle:** [◇ 3 Point Corner Rectangle] A *3 Point Corner Rectangle* is defined by three corner points. This allows you to specify the rectangle's orientation.
- **3 Point Center Rectangle:** [◈ 3 Point Center Rectangle] A *3 Point Center Rectangle* is defined by a center point, a corner point, and a side midpoint. This allows you to specify the rectangle's orientation.
- **Parallelogram:** [▱ Parallelogram] A *Parallelogram* is defined by three corner points.

1.4.5) Circles

You are able to draw different circle types. The circle commands are located in the **Sketch** tab in the *Command Manager*. A description of each circle type is given below.

- **Circle:** ⊘ Circle Created by selecting a center point and then a point on its circumference.
- **Perimeter Circle:** ⊕ Perimeter Circle Created by selecting 3 points on its circumference.

1.4.6) Arcs

You are able to draw different arc types. The arc commands are located in the **Sketch** tab in the *Command Manager*. A description of each arc type is given below.

- **Centerpoint Arc** ⟳ Centerpoint Arc : Defined by a center point, a starting point and an ending point.
- **3 Point Arc** ⌒ 3 Point Arc : Defined by a start point, endpoint and a radius.
- **Tangent Arc** ⤵ Tangent Arc : Defined by the endpoint of another entity that it will be tangent to and an endpoint.

1.4.7) Slots

You are able to draw different slot types. The slot commands are located in the **Sketch** tab in the *Command Manager*. A description of each slot type is given below.

- **Straight Slot** ⊙⊙ Straight Slot : Sketches a straight slot using two end points.
- **Centerpoint Straight Slot** ⊙ Centerpoint Straight Slot : Sketches a straight slot from the center point.
- **3 Point Arc Slot** 3 Point Arc Slot : Sketches an arc slot using three points along the arc.
- **Centerpoint Arc Slot** Centerpoint Arc Slot : Sketches an arc slot using the center point of the arc radius and the two end points.

1.4.8) Polygon

The **Polygon** 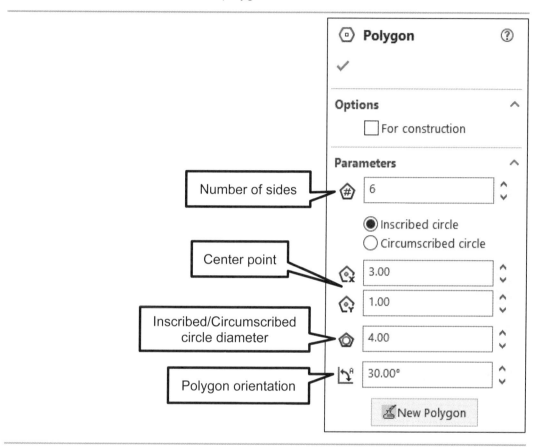 command is located in the *Sketch* tab in the *Command Manager*. The Polygon options window is shown in Figure 1.4-2 and Figure 1.4-3 shows the difference between an inscribed polygon and a circumscribed one.

Figure 1.4-2: Polygon options window

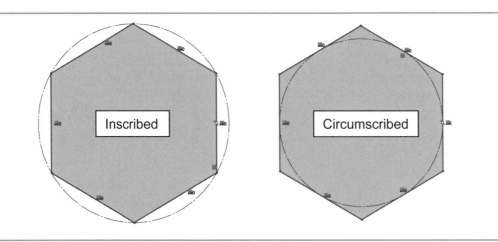

Figure 1.4-3: Inscribed and Circumscribed Polygons

1.4.9) Sketch Chamfer/Fillet

The **Sketch Chamfer** [Sketch Chamfer] and **Sketch Fillet** [Sketch Fillet] commands are located in the *Sketch* tab in the *Command Manager*. A *Chamfer* is a beveled corner. It can be defined by selecting a vertex or two edges and then specifying a distance and an angle, or two distances. A *Fillet* is a rounded corner. It is defined by selecting a vertex or two edges and then specifying a radius. You should set your fillet or chamfer size before you select the entity. If a warning window appears, it means that by applying the chamfer or fillet you will be deleting a sketch relation or dimension. You may need to reapply the deleted constraint.

It is important to understand that these are chamfers and fillets that are created in a sketch. There are also feature chamfer and fillet commands that are performed on a solid. These will be discussed later. This *Sketch Chamfer* and *Sketch Fillet* options windows are shown in Figure 1.4-4.

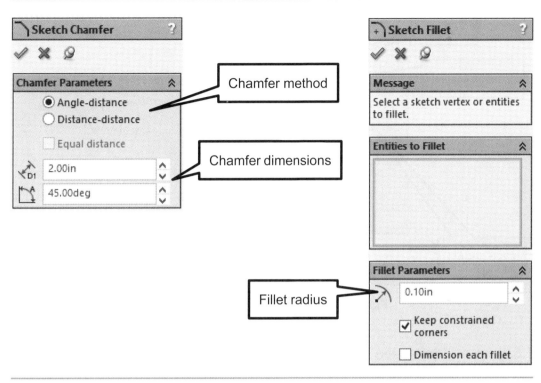

Figure 1.4-4: Sketch Chamfer and Fillet option windows

1.5) DIMENSIONS AND RELATIONS

1.5.1) Sketch relations

Sketch relations add geometric constraints between two or more entities. For example, we can make two lines parallel, or two circles concentric. To apply sketch relations between entities, follow the steps below. Once a relation has been applied, you should see a sketch relation symbol indicating which relation has been applied. For example, if two lines are perpendicular, you will see this symbol [⊥]. If the sketch relations do not appear, you can view sketch relations by selecting **View – Hide/Show - Sketch Relations** from the pull-down menu. Figure 1.5-1 shows some examples of applied sketch relations.

<u>Applying a sketch relation</u>

1) Select one of the elements that you want to apply the relation to.
2) Hold the **Ctrl** key and then select the next element where you want to apply the relation.
3) Continue selecting elements if you want to apply the relation to more than two elements.
4) In the *Properties* window, select the relation that you wish to apply.

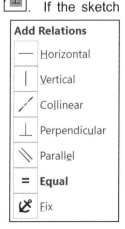

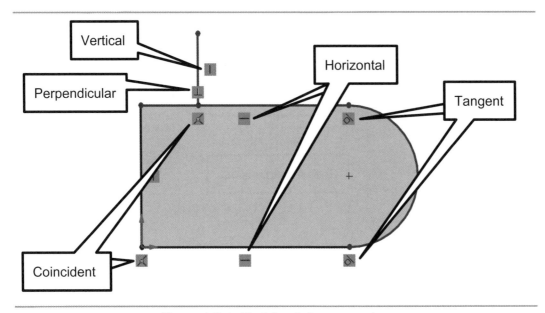

Figure 1.5-1: Sketch relation examples

1.5.2) Smart Dimension

When you sketch in SOLIDWORKS®, you are generally going to sketch the approximate shape first. Then, after the general shape has been created, you will add sketch relations and dimensions. The smart dimension command will generate the type of dimension the program believes you want to use. It is usually correct. There are ways to change the type of dimension if it is chosen incorrectly, but this is rare. To apply a dimension, use the following steps. The *Modify* and *Dimension* options windows are shown in Figure 1.5-2.

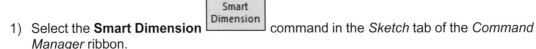

1) Select the **Smart Dimension** command in the *Sketch* tab of the *Command Manager* ribbon.
2) Select the feature(s) that you want to dimension or dimension between.
3) Pull the dimension out and place it by clicking your left mouse button.
4) Within the *Modify* or *Dimension* options window, fill in the correct dimension value and then select the green check mark ✓.

1.5.3) Editing a Dimension

You can edit a dimension after it has been entered. To edit a dimension, double-click on the numerical value of the window which will bring up the *Modify* window. Or, single-click anywhere on the dimension to bring up the *Dimension* options window. Figure 1.5-2 shows both the *Modify* and *Dimension* options windows.

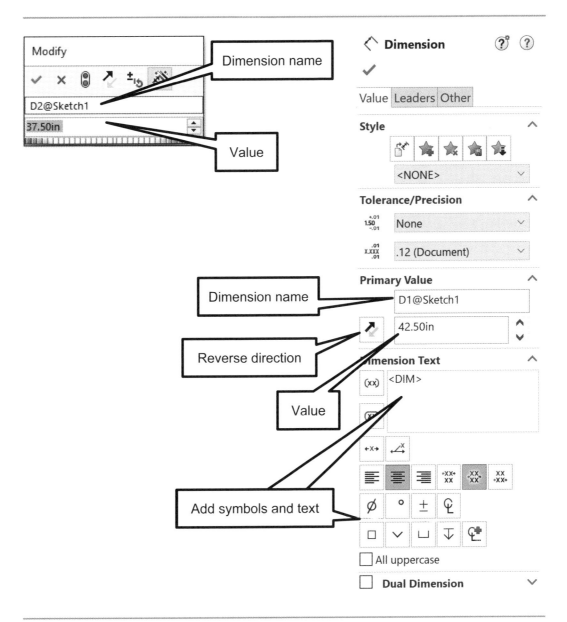

Figure 1.5-2: *Dimension* option window

1.6) FEATURES

Once a sketch has been created, a feature can be applied. A feature creates a solid based on your sketch. The feature commands are located in the *Feature* tab shown in Figure 1.6-1. To apply a feature command to a sketch you can select the feature command while you are still in the sketch. The program will automatically exit you out of the sketch. Or you can exit the sketch, select the feature command, and select the sketch where you want the command applied.

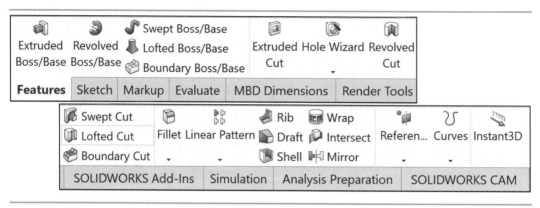

Figure 1.6-1: Feature tab commands

1.6.1) Extrude and Extrude cut

Extrude Boss/Base and **Extruded Cut** are the most commonly used features commands. The *Extrude Boss/Base* and *Extrude Cut* commands are essentially applied in the same way. The only difference is that the *Extrude Boss/Base* command adds material and the *Extrude Cut* command removes material. There are several options you can select when applying the extrude commands. To learn about these options, see Figure 1.6-2. The *Extrude Boss/Base* and *Extrude Cut* commands allow you to select between several different extrude methods. There is also an option to create a *Thin Extrude*. This may be done with an open or closed profile. Figure 1.6-3 shows some examples of the different *Extrude* methods.

Extrude methods

- Blind: Input a specific distance.
- Through All: Extrude through all material.
- Up to Vertex: Select a vertex you wish to extrude up to.
- Up to Surface: Select a surface you wish to extrude up to.
- Offset from Surface: Select a surface that you wish to extrude up to and then an offset from that surface.
- Up to Body: Select a body you wish to extrude up to.
- Mid Plane: Extrude about the mid plane. The extrude distance is the total distance. Half of the distance on each side of the mid plane.

Addition methods for *Extruded Cut*

- <u>Through All:</u> Extrude through all material.
- <u>Through – Both:</u> Extrude through all material in both directions.

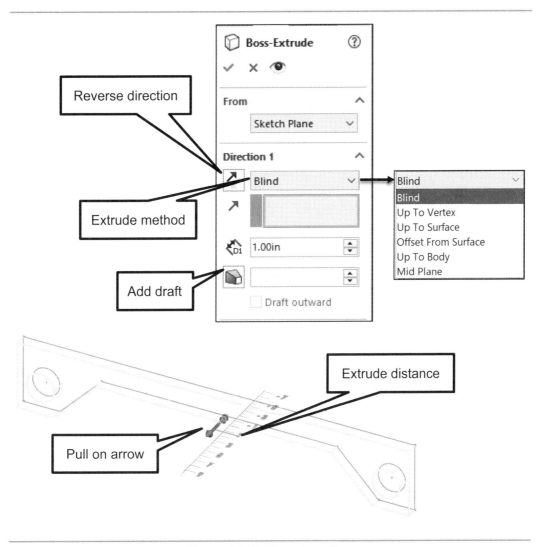

Figure 1.6-2: Extrude options

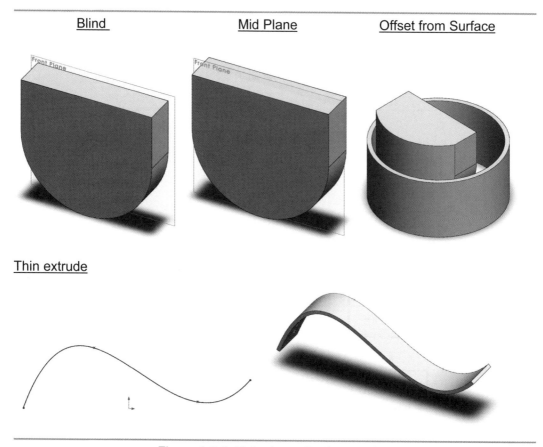

Figure 1.6-3: Extrude method examples

1.6.2) Fillet and Chamfer

The *Fillet* and *Chamfer* commands are similar to the *Sketch Fillet* and *Sketch Chamfer* commands previously described. The difference is that these commands are applied to a solid and not a sketch. Figure 1.6-4 shows the *Fillet* and *Chamfer* option windows.

- **Fillet:** A *Fillet* is a rounded corner. It is created by selecting two faces or an edge and then specifying its radius. You may apply several fillets at once. You can also specify the fillet type and profile.

- **Chamfer:** A *Chamfer* is a beveled corner. It is created by selecting two faces or an edge and then specifying either a distance and angle or two distances. You may apply several chamfers at once.

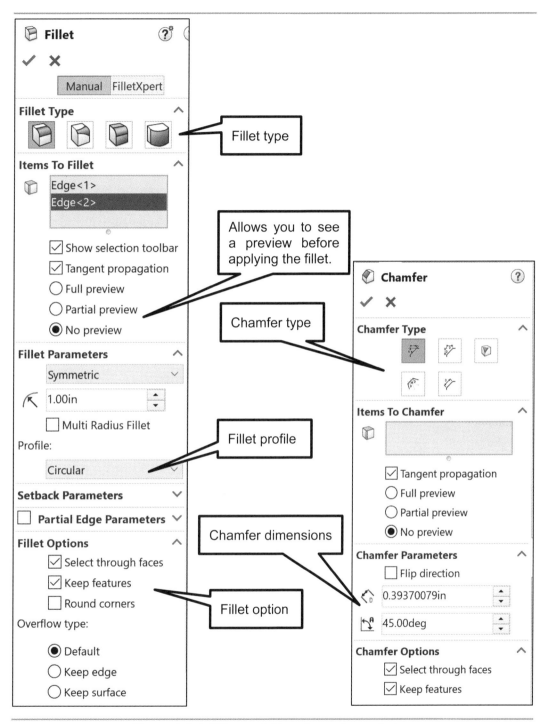

Figure 1.6-4: Fillet and Chamfer options window

1.7) FEATURE MANAGER DESIGN TREE

The *Feature Manager Design Tree*, or the *Feature Tree* for short, contains all the sketches, features and operations that you have taken during the creation of your part. Figure 1.7-1 shows the part that you will be creating in your first tutorial and the associated feature tree. First a circle sketch1 was used to create the first Boss-Extrude1 or cylinder and then sketch2 was used to create the second or smaller Boss-Extrude2. Lastly, sketch3 was used to Cut-Extrude1 the hole in the part.

The Feature Tree allows you to go back and edit sketches and features, reorder the features, and to roll-back the steps. To learn more about these topics see the information boxes on the following pages.

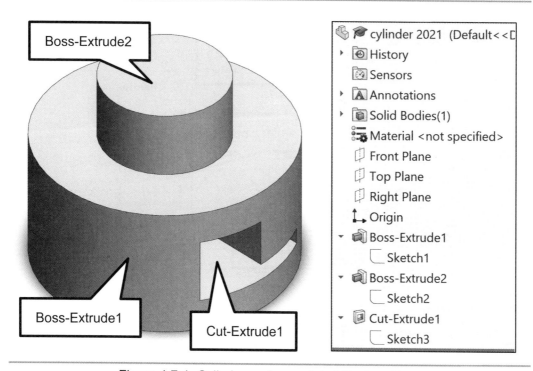

Figure 1.7-1: Cylinder and associated Feature Tree

1.7.1) Editing a Feature

You can edit a feature after it has been applied using the following steps.

1) Select the feature to be edited in the *Feature Manager Design Tree*. Right clicking shows you the full command list.
2) Select **Edit Feature** from the *Context* toolbar.

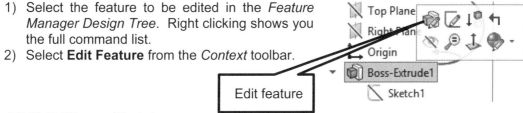

1.7.2) Editing a Sketch

You can edit a sketch after it has been created and after a feature has been applied by using the following steps.

1) Select the sketch to be edited in the *Feature Manager Design Tree*.

2) Select **Edit Sketch** from the *Context* toolbar.

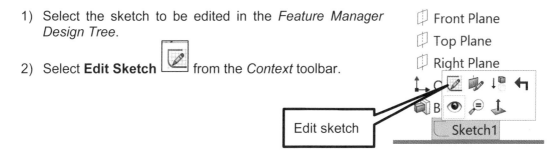

1.7.3) Rollback Bar

You can use the *Feature Manager Design Tree* to temporarily roll back to an earlier state. You can then add new features or edit existing features while the model is in the rolled-back state. You can save the model with the rollback bar in any place. To roll back your model, grab the rollback bar. A hand should appear. Drag the bar up or down depending on your desired rollback direction. See Figure 1.7-2 to see a part before and after a rollback.

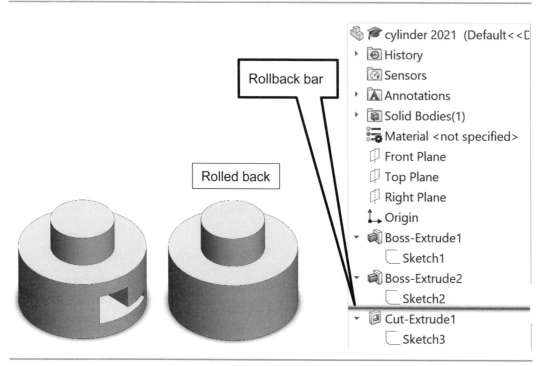

Figure 1.7-2: Using the rollback bar

1.8) VIEWING YOUR PART

When you are modeling your part or assembly, it is important to be able to view it from many different angles. A few very useful shortcut keys for manipulating how you view your part are:

- **F**: Zooms to fit.
- **Z**: Zooms out.
- **Shift+Z**: Zooms in.
- **Ctrl+8**: Shows your sketching plane straight on.
- **Ctrl+7**: Shows an isometric view of your part.
- **Ctrl+drag the middle mouse button**: Pans the model.
- **Ctrl+arrow keys:** Pans the model.
- **Drag the middle mouse button:** Rotates the model.

Another way to select the direction that you want to view your model is to use the *View Selector*. When you hit the **Space bar** the *View Selector* and an *Orientation* window will appear. Figure 1.8-1 shows the *View Selector* and *Orientation* window. View orientations may also be accessed from the *Heads-up* toolbar. You can also access the zoom command here. The *Heads-up* toolbar is shown in Figure 1.8-2.

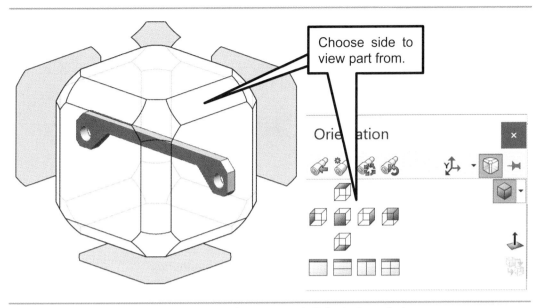

Choose side to view part from.

Figure 1.8-1: *View Selector* and *Orientation* window

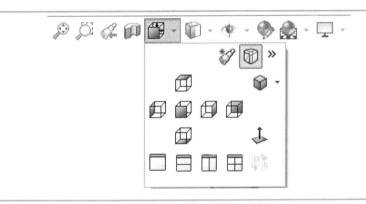

Figure 1.8-2: *Heads-up* toolbar

1.9) ASSIGNING MATERIAL

You can assign a material to your model at any stage of the design process. Adding a material allows you to calculate the mass, center of gravity, and perform simulations on your part. Use the following steps to assign a material to your part. The *Material* options window is shown in Figure 1.9-1.

1) Right click on **Material** in the *Feature Manager Design Tree*.
2) Select **Edit Material**.
3) A *Material* window will appear.
4) Select your desired material.
5) Select **Apply**.
6) Select **Close**.

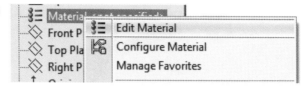

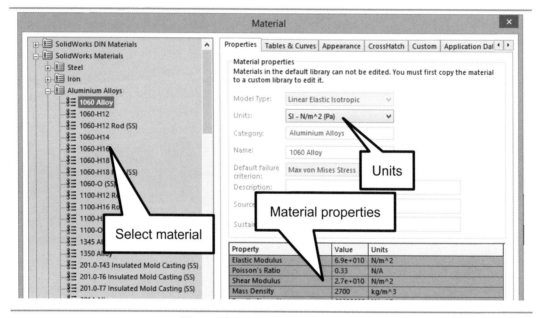

Figure 1.9-1: *Heads-up* toolbar

1.10) EVALUATE

The commands located in the **Evaluate** tab (shown in Figure 1.10-1) allow you to calculate the mass and center of gravity of your model, measure distances, calculate interferences, apply sensors, and other commands that will evaluate the model. We will talk about some of these in upcoming chapters. In this chapter we will focus on the **Mass Properties** command.

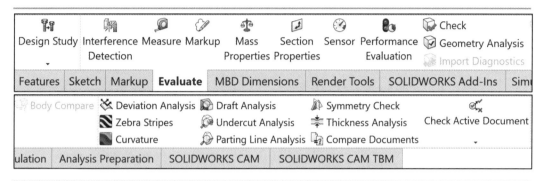

Figure 1.10-1: Evaluate tab commands

The **Mass Properties** command calculates the following about your model. The *Mass Properties* window is shown in Figure 1.10-2.

- Mass (weight if you are using US customary units)
- Volume
- Surface area
- Center of mass
- Mass moment of inertia

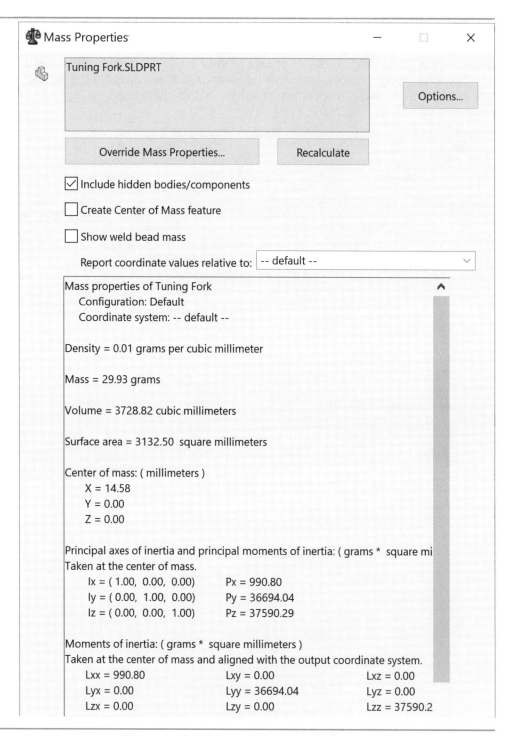

Figure 1.10-2: Mass Properties window

1.11) CYLINDER TUTORIAL

1.11.1) Prerequisites

To complete this tutorial, you are not expected to have any prior experience with solid modeling programs. The following topics are prerequisites for this tutorial.

- Familiarity with computer navigation (e.g. how to open a program, what is a pull-down menu).
- How to read dimensions.

1.11.2) What you will learn?

The objective of this tutorial is to familiarize the user with the SOLIDWORKS® modeling basics. This tutorial will cover the basic modeling command sequences used throughout the program. It will implement tools such as sketching and dimensioning, as well as extrusions and extrude cuts. You will be modeling the object shown in Figure 1.11-1.

File setup and User interface

- New
- Save
- Help
- Resources
- Pull-down menu
- Units
- Viewing the sketch plane

Sketcher

- Sketching on a plane or face.
- Circle
- Line

Features

- Extrude Base/Boss
- Extrude Cut

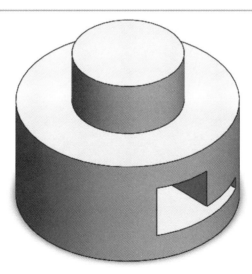

Figure 1.11-1: Cylinder model

1.11.3) Setting up the part

1) **Start SOLIDWORKS**

2) If a *Welcome – SOLIDWORKS* window appears, notice that, in the *Learn* tab, there is SOLIDWORKS information, tutorials, and training. **Close** this window.

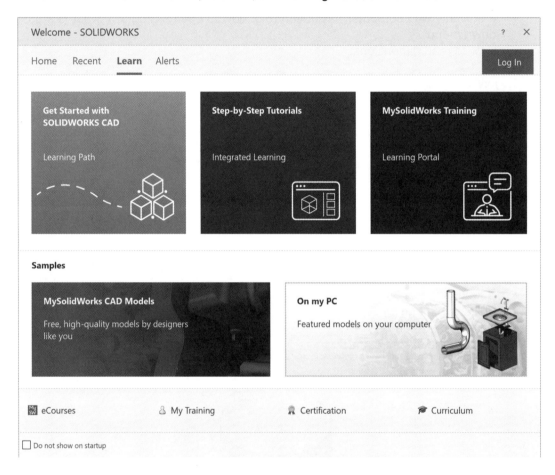

3) Click on the **Help** icon (located at the top right) and then click on **Help** again. This will take you to a help site. This site allows you to search for specific 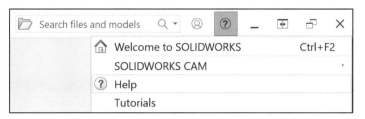 commands and retrieve information explaining the uses of the command. Note that to the left of the *Help* icon is the place where you can search commands as well. After looking at the help section, leave this area and go back to SOLIDWORKS.

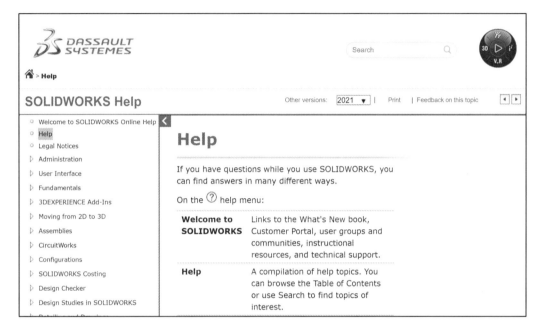

4) Pin the pull-down menu. The pull-down menus may be accessed by clicking on the black triangle just to the right of the SOLIDWORKS logo in the top left corner. You can pin the pull-down menu by clicking on the push pin icon.

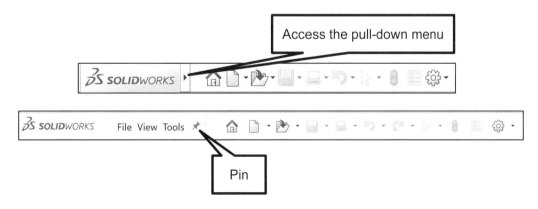

5) In the *Quick Access* toolbar (located at the top), select **new** and then select **part**

.

> ➤ Note: If a warning window appears that refers to a file, click **Yes**.
> ➤ Note: If a warning window appears that refers to a default template, click **OK**.
---

6) Notice the user interface for your new part. Important features are pointed out in the figure below.

> ➤ See section 1.3 on *New Part, User Interface,* and *Set Up* for more information about each area.

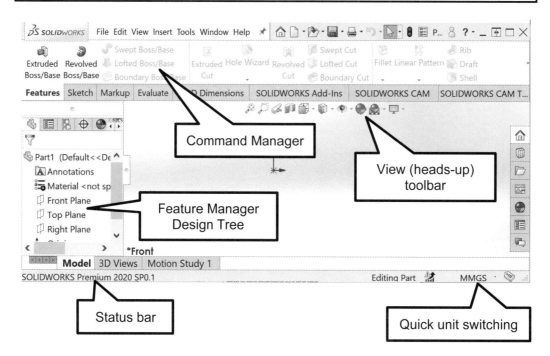

7) Set your drafting standard to **ANSI** and set the text to **upper case** for notes, tables, and dimensions.

    a. Select the **Options** icon (located at the top): ⚙

    b. Click on the **Document Properties** tab.

    c. Click the **Drafting Standards** option in the left column.

    d. Select the **ANSI** drafting standard from the pull-down menu at the top right (under Overall drafting standard).

    e. Check all categories for **uppercase** lettering. When you do this, your drafting standard will change to MODIFIED.

    f. Select **OK**.

System Options	Document Properties
Drafting Standard	Overall drafting standard
⊞ Annotations	ANSI-MODIFIED
⊞ Dimensions	
Virtual Sharps	Derived from: ANSI
⊞ Tables	
⊞ DimXpert	
Detailing	
Grid/Snap	Uppercase
Units	
Model Display	☑ All uppercase for notes
Material Properties	
Image Quality	☑ All uppercase for tables
Sheet Metal MBD	
Sheet Metal	☑ All uppercase for dimensions and hole callouts

8) Set your unit to **IPS** (i.e., inch, pound, second) and set your **Decimals** to **.12**. Also, select the rounding option to **Round half to even**.

    a. Select the **Options** icon (located at the top):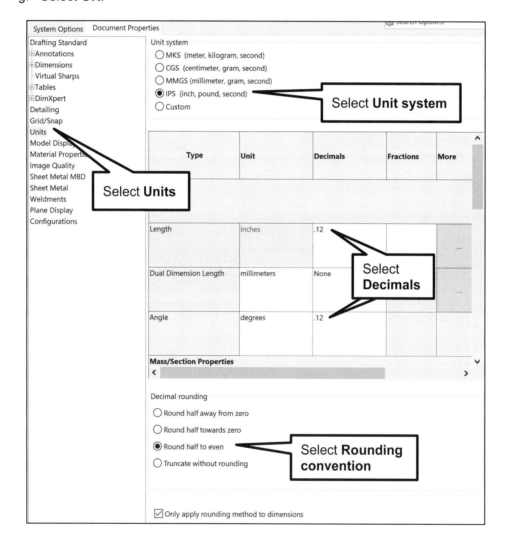
    b. In the same **Document Properties** tab
    c. Click on **Units** in the left column.
    d. Select **IPS** as your *Unit system*.
    e. Set your *Decimals* to **.12**. (See figure below)
    f. Select **Round half to even** as your *Decimal rounding*.
    g. Select **OK**.

9) **Save** 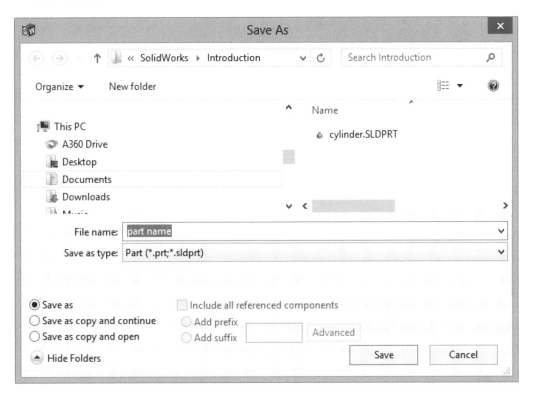 your part as **CYLINDER.SLDPRT**. Remember to save often throughout this tutorial.

1.11.4) Sketching and Extrude Boss/Base

1) In the *Sketch* tab, select the **Sketch** command and then select the **Top Plane** in the drawing area.

2) Sketch a **Circle** whose center is **Coincident** with the origin and diameter is **2 inches**.

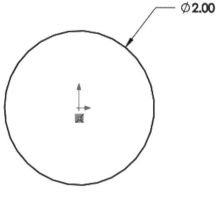

Ø**2.00**

a. Select the **Circle** 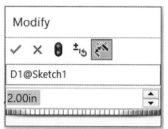 command.

b. Bring the cursor to the origin and an orange snap will appear, **click** and bring your cursor outward and **click** again to finalize the circle.

c. Select the **Smart Dimension**
```
        ⌒
Smart Dimension
```
command. Hover your cursor on the perimeter of the circle and click. Pull away from the circle to see the diameter dimension and click. Enter **2** in the *Modify* window and select ✓.

Modify
✓ × ⬮ ±↺ ⚒
D1@Sketch1
2.00in

> ➢ See section 1.4 on **Sketching** for more information.
> ➢ See section 1.5.2 on **Smart Dimensioning** for more information.

3) In the *Feature* tab, select the **Extrude** ⬚ Extruded Boss/Base command, select **Blind** as the type of extrude and enter **1.00** inch as the distance in the *Boss-Extrude* window and click the green checkmark.

> ➢ See section 1.6.1 on **Extrude** and **Extrude Cuts** for more information.

4) Click on the top surface and then select **Sketch** 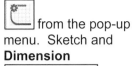 from the pop-up menu. Sketch and **Dimension**  a **1.00 inch** diameter **Circle** on the top face of the cylinder. Make the center of the circle **Coincident** with the origin.

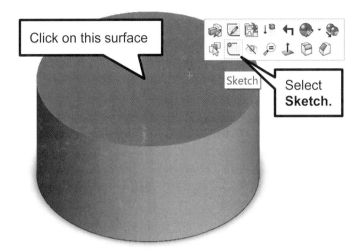

Click on this surface

Sketch

Select **Sketch.**

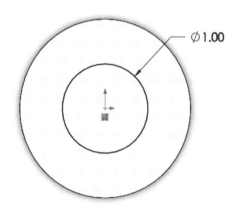

⌀1.00

5) **Extrude** the circle to a distance of **0.50 inch**.

Boss-Extrude2

From

Sketch Plane

Direction 1

Blind

0.50in

☑ Merge result

1.11.5) Sketching and extrude cut

1) In the *Feature Manager Design Tree*, select the Right Plane. Then, in the **Sketch tab**, select the **Sketch** command.

2) Hit **Ctrl+8**. This will show you the normal view.

3) **Sketch** and **Dimension** the profile shown using the **Line** command. Sketch the shape first. Then add dimensions. Note that after you have completely dimensioned your sketch, it will turn **black** indicating that it is completely constrained.

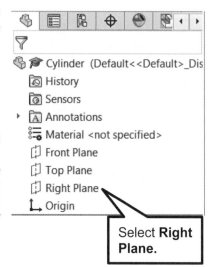

Select **Right Plane.**

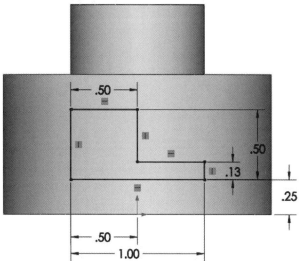

> ➢ See section 1.4 on **Sketching** for more information.
> ➢ See section 1.5.2 on **Smart Dimensioning** for more information.

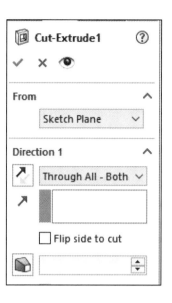

4) **Extruded Cut** the sketch using the **Through All – Both** method. Select **Ctrl + 7** on the keyboard to enter the isometric view.

5) **Save** your part.

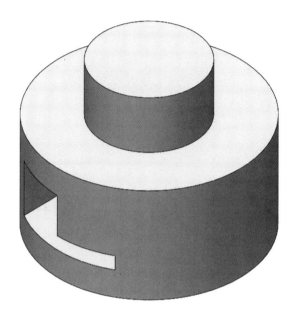

➢   See section 1.6.1 on **Extrude** and **Extrude Cuts** for more information.

1.12) ANGLED BLOCK TUTORIAL

1.12.1) Prerequisites

Before completing this tutorial, you should have completed the following tutorial and be familiar with the following topics.

Pre-requisite Tutorial

- Chapter 1 – Cylinder Tutorial

Pre-requisite Topics

- Computer navigation.
- Passing familiarity with orthographic projection.
- Ability to read dimensions.

1.12.2) What you will learn?

The objective of this tutorial is to introduce you to creating simple *Sketches*, *Extrudes* and *Cuts*. You will be modeling the *Angled Block* shown in Figure 1.12-1. Specifically, you will learn the following commands and concepts.

Sketching

- Rectangle
- Line
- Circle
- Smart dimension

Features

- Extrude
- Extrude Cut

Material and properties

- Applying material
- Mass properties

View

- Viewing the sketch plane

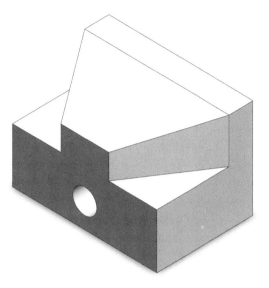

Figure 1.12-1: Angled Block Model

1.12.3) Setting up the project

1) **Start SOLIDWORKS** and then start a **new** 📄 **part** .

2) Set your drafting standard to **ANSI** and set all your text to **upper case** for notes, tables and dimensions. (*Options* ⚙ *– Document Properties – Drafting Standard*)

3) Set your unit to **IPS** (i.e., inch, pound, second) and set your **Decimals = .12**. Also, select the rounding option- **Round half to even**. (*Options* ⚙ *– Document Properties – Units*)

4) Save your part as **ANGLE BLOCK.SLDPRT** (**File – Save**). Remember to save often throughout this project.

1.12.4) Creating the part

1) **Sketch** ⌐ *Sketch* on the **Front Plane**.

2) Draw a **Rectangle** that has one corner coincident with the origin

and **Dimension** 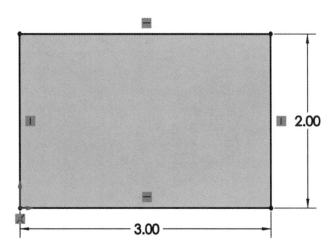 it as shown in the figure.

> ➢ See section 1.4 on *Sketching* for more information.
> ➢ See section 1.5.2 on *Smart Dimensioning* for more information.

3) **Extrude** the rectangle **2** inches.

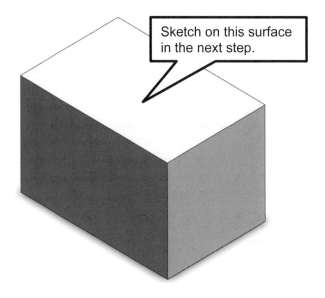

Sketch on this surface in the next step.

4) **Sketch** 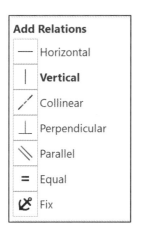 on the top surface of the block and create the sketch shown using **Lines** . Use **Ctrl+8** to see the normal view. Use the **Equal** sketch relation to constrain the left side sketch to the right side sketch. To apply the sketch relation, hold the **Ctrl** key and click on the two lines you want to make equal in length. Then, select **Equal** in the *Properties* window. If your sketch relations are not showing up on your sketch, select **View – Hide/Show – Sketch Relations** from the top pull-down menu. Your sketch should be completely black when done.

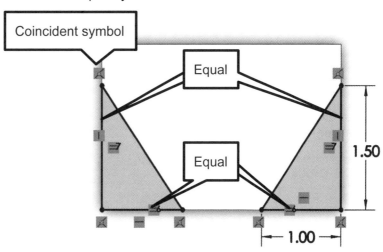

> ➢ To learn more about **Sketch Relations** see section 1.5.1.

5) **Cut Extrude** the sketch **0.9** inches using the **Blind** method. Hit **Ctrl+7** to view the isometric.

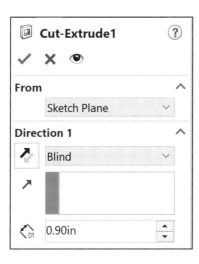

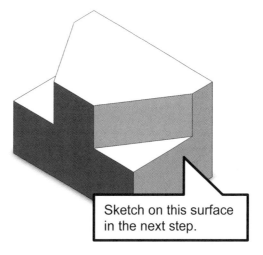

Sketch on this surface in the next step.

6) **Sketch** 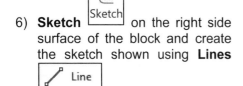 on the right side surface of the block and create the sketch shown using **Lines** 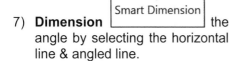.

7) **Dimension**  the angle by selecting the horizontal line & angled line.

8) **Cut Extrude** the sketch **Through All**.

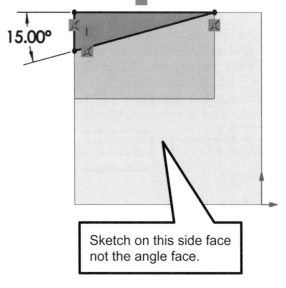

15.00°

Sketch on this side face not the angle face.

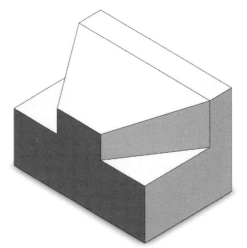

9) **Sketch** on the front face of the block and create the sketch shown using a **Circle**

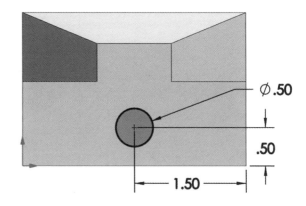

Ø.50

.50

1.50

10) **Cut Extrude** the sketch **Through All**.

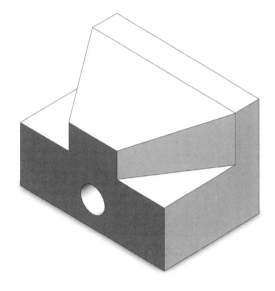

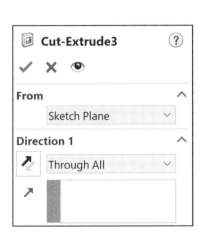

1.12.5) Adding material

1) In the *Feature Manager Design Tree*, **right-click** on **Material** and select **Edit Material**.

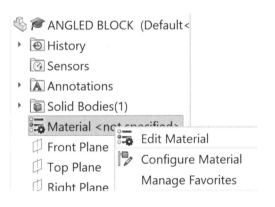

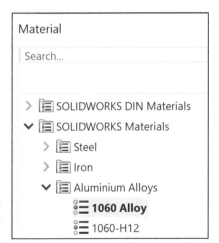

2) In the *Material* window, expand *SOLIDWORKS Materials* and *Aluminum Alloys*. Then, select **1060 Alloy**. Select **Apply** and then **Close**.

> ➢   To learn more about ***Adding Material*** see section 1.9.

3) Calculate the weight of your part. In the **Evaluate** tab, select **Mass Properties**

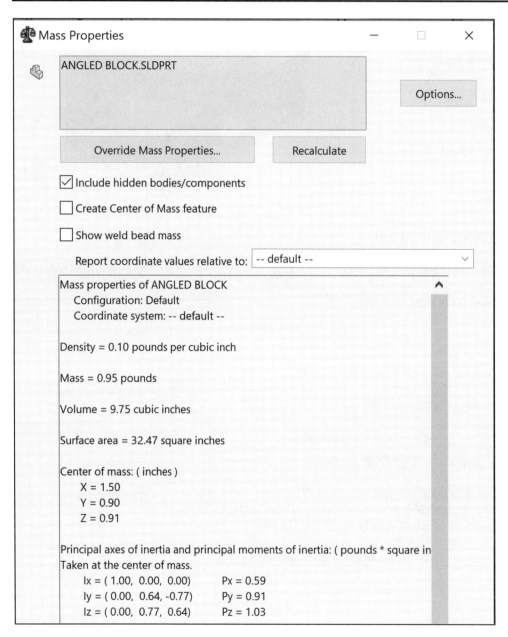

. In the *Mass Properties* window note that the mass of your part is 0.95 lb. This is the actual weight of your part because of the units. If your weight is not this value, your model is incorrect. This window also gives other physical properties.

> ➢ To learn more about the **Mass Properties** command, see section 1.10.

Mass Properties   —   ☐   ✕

ANGLED BLOCK.SLDPRT

Options...

Override Mass Properties...   Recalculate

☑ Include hidden bodies/components

☐ Create Center of Mass feature

☐ Show weld bead mass

Report coordinate values relative to: -- default --   ⌄

Mass properties of ANGLED BLOCK   ⌃
   Configuration: Default
   Coordinate system: -- default --

Density = 0.10 pounds per cubic inch

Mass = 0.95 pounds

Volume = 9.75 cubic inches

Surface area = 32.47 square inches

Center of mass: (inches)
   X = 1.50
   Y = 0.90
   Z = 0.91

Principal axes of inertia and principal moments of inertia: (pounds * square in
Taken at the center of mass.
   Ix = (1.00, 0.00, 0.00)   Px = 0.59
   Iy = (0.00, 0.64, -0.77)   Py = 0.91
   Iz = (0.00, 0.77, 0.64)   Pz = 1.03

---

4) **IMPORTANT! Save** your part; this model will be used in a future tutorial.

1.13) CONNECTING ROD TUTORIAL

1.13.1) Prerequisites

Before completing this tutorial, you should have completed the following tutorial and be familiar with the following topics.

Pre-requisite Tutorial

- Chapter 1 – Angled Block Tutorial

Pre-requisite Topics

- Computer navigation.
- Passing familiarity with orthographic projection.
- Ability to read dimensions.

1.13.2) What you will learn

The objective of this tutorial is to introduce you to creating simple *Sketches*, *Extrudes* and *Cuts*. You will be modeling the *Connecting Rod* shown in Figure 1.13-1. Specifically, you will learn the following commands and concepts.

Sketching

- Sketch relations
- Editing dimensions
- Editing sketches
- Sketch chamfers
- Sketch fillet
- Rectangle

Features

- Chamfer
- Fillets
- Editing a feature

Material and properties

- Applying material
- Mass properties

View

- Panning
- Rotating

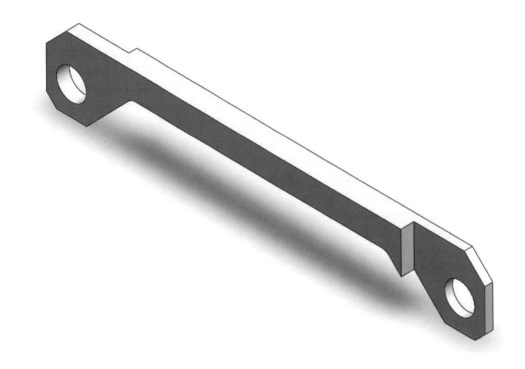

Figure 1.13-1: Connecting rod Model

1.13.3) Setting up the project

1) **Start SOLIDWORKS** and then start a **new** 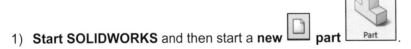 **part** .

2) Set your drafting standard to **ANSI** and set all the text to **upper case** for notes, tables, and dimensions. (*Options* ⚙ *– Document Properties – Drafting Standard*)

3) Set your unit to **IPS** (i.e., inch, pound, second) and set your **Decimals = .12**. Also, select the rounding option- **Round half to even**. (*Options* ⚙ *– Document Properties – Units*)

4) **Save** your part as **CONNECTING ROD.SLDPRT** (**File – Save**). Remember to save often throughout this project.

1.13.4) Base extrude

1) **Sketch**  on the **Front Plane**.

2) Use the **Circle**  command to sketch two circles as shown below. Make one of the circle centers **coincident** with the origin. You will know when you have snapped to the origin when a small circle appears. Don't worry about the circle's spacing or size at this point. Just make them about 5 or more diameters apart.

3) **Pan** your drawing area to center the circles by holding down the **Ctrl** key and the **middle mouse button** and then moving the mouse.

> ➢ For more information on *Viewing Your Part* see section 1.8.

4) Use the **Line** command to sketch the following profile. Be approximate. Don't worry about getting it exact. Notice that when you are drawing the lines, dashed lines will occasionally appear. These dashed lines allow you to snap to the origin or to the geometric features of the object that have already been drawn.

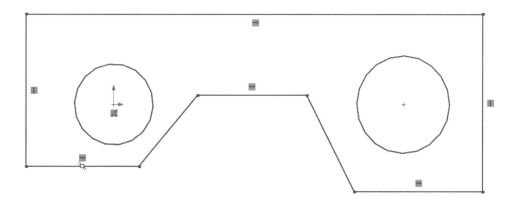

5) View your sketch relations (**View – Hide/Show – Sketch Relations**) if they are not already visible. Your sketch relations will show up as symbols inside a green box.

6) Add the following sketch relations. Don't worry if your drawing goes wonky. Just click and drag the elements into position. Remember to use the **Ctrl** key to multiple select.

Note: To add a *Sketch Relation*, click on the element(s) and select the relation in the *Properties* window on the left.

> ➢   To learn more about ***Sketch Relations*** see section 1.5.1.

---

    a) If any of the **horizontal** or **vertical** lines are not perfectly horizontal or vertical, add those relations.
    b) Make the two circle diameters **Equal** (click on circumference of each circle).
    c) Make the circle centers **Horizontal** (click on the circle centers).
    d) Make the two bottom horizontal lines **Collinear**.
    e) Make the two bottom horizontal line lengths **Equal**.
    f) Make the two angled line lengths **Equal**.

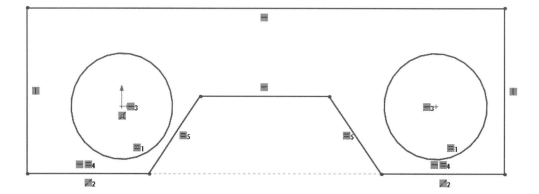

7) Add the **Dimensions** 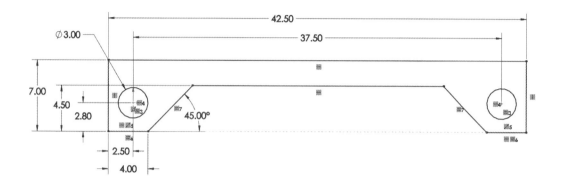 shown. If your drawing exceeds your viewing area, select the **F** key to fit all. When dimensioning, it is a good idea to start with the overall dimensions and then work down to the smaller dimensions.

Note: If a ***Make Dimension Driven?*** window appears, you have an unwanted sketch relation or you have a duplicate dimension. Select **Cancel** and then search and delete the extra constraint.
Note: If, while you are dimensioning the part, the **angled lines become parallel**, you need to remove the equal constraint and then reapply it after you have adjusted the lines.

> ➤   To learn more about ***Dimensioning*** see section 1.5.2.

8) After you are done dimensioning, there should be **no blue lines**. Blue lines mean that it is under-constrained. If you have blue lines, try the following.
   a) Make sure that the left circle and the origin are **Coincident**.
   b) Click and pull on the blue element. It will move in the direction that it is not constrained. Add the appropriate constraint.

9) **Extrude** 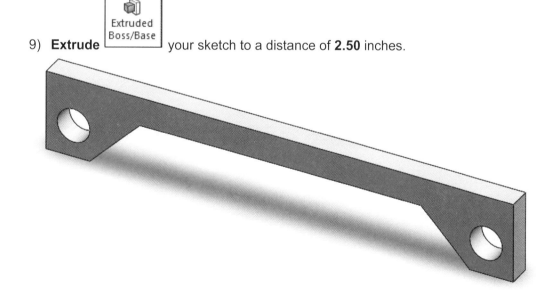 your sketch to a distance of **2.50** inches.

10) Try **zooming in and out** by scrolling your middle mouse wheel. Notice that the mouse location identifies the zooming center.

11) Fit all (**F**).

> ➤ For more information on **_Viewing Your Part_** see section 1.8.

12) **Edit** your **sketch**. To do this, expand your *Boss-Extrude1* by clicking on the black arrow. Then click on the sketch and select **Edit Sketch**.

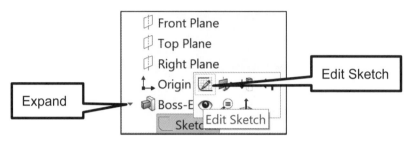

> ➤ For more information on **_Editing a Sketch_** see section 1.7.2.

13) View your sketch from the normal plane by selecting **Ctrl + 8**.

14) Add two **2 x 45° Chamfers** 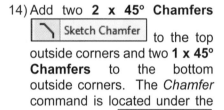 to the top outside corners and two **1 x 45° Chamfers** to the bottom outside corners. The *Chamfer* command is located under the *Fillet* 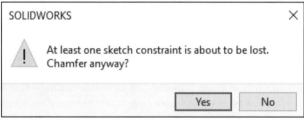 command. When applying the 1 x 45°, select **Yes** in the warning window. Note that this will delete the *Equal* relation between the two bottom horizontal lines. Reapply the **Equal** relation after you have applied the chamfer. After you reapply the *Equal* relation, you may be asked to **Rebuild and Save**.

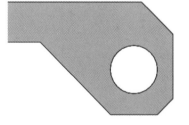

> ➤ For more information on **_Sketch Chamfers and Sketch Fillets_** see section 1.4.9.

15) **Exit Sketch** and view your part from the isometric view (**Ctrl + 7**).

16) Change the **Extrude** distance from 2.5 to **2.1 inches**. To do this, click on your *Boss-Extrude* and select **Edit Feature**.

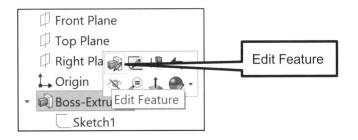

> ➢ For more information on *Editing a Feature* see section 1.7.1.

1.13.5) Adding features

1) **Sketch** on the top face of your part.

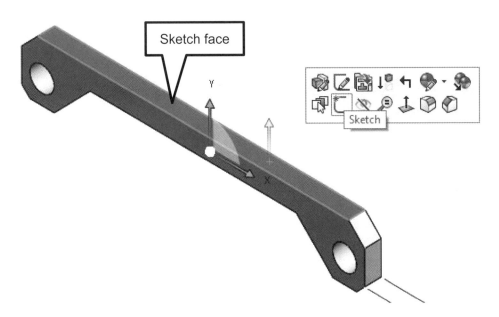

2) View the sketch from the normal direction (**Ctrl + 8**).

3) Sketch and dimension the following two **Rectangles** [Corner Rectangle].

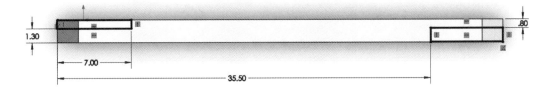

4) **Extrude Cut** the rectangles **Through All**.

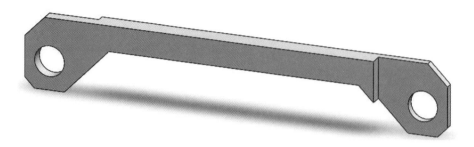

5) Add **R1.00 Fillets** [Fillet] to the area where the angled lines meet the main body of the rod (*Feature* tab). You may need to **rotate** your part to view the underside of the part. To do this, click the **middle mouse button** and move your mouse. To apply the fillet, select the edge where the fillet will be applied.

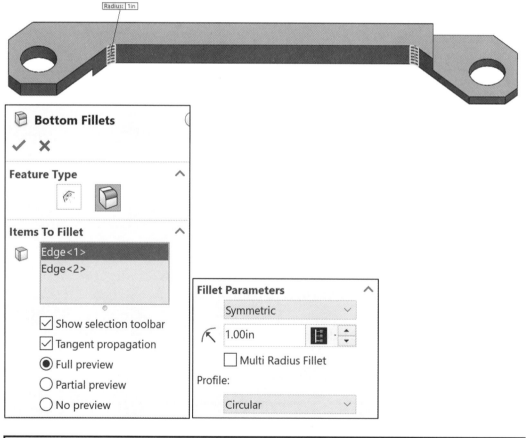

> ➢ For more information on *Fillets* see section 1.6.2.

6) In your *Feature Manager Design Tree*, name your features as shown. To name your feature, slowly double click on the name and type in the new name.

7) **Save**.

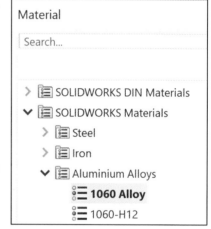

CONNECTING ROD
▸ Solid Bodies(1)
▸ Annotations
▸ Equations
  1060 Alloy
  Front Plane
  Top Plane
  Right Plane
  Origin
▸ Base
▸ Side Cuts
  Bottom Fillets

1.13.6) Adding material

1) Apply a material of **Aluminum 1060 Alloy** to your part.

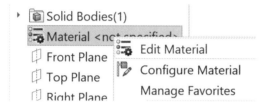

▸ Solid Bodies(1)
  Material <not specified>
  Front Plane   Edit Material
  Top Plane   Configure Material
  Right Plane   Manage Favorites

2) Calculate the weight of your part. In the **Evaluate** tab, select **Mass Properties**

. In the *Mass Properties* window, note that the mass of your part is 22.22 lb. This is the actual weight of your part because of the units. If your weight is not this value, your model is incorrect. This window also gives other physical properties.

Material

Search...

> SOLIDWORKS DIN Materials
✓ SOLIDWORKS Materials
  > Steel
  > Iron
  ✓ Aluminium Alloys
    1060 Alloy
    1060-H12

3) **IMPORTANT! Save** and **keep** your part. This model will be used in a future tutorial.

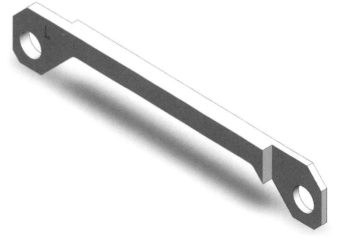

NOTES:

BASIC MODELING IN SOLIDWORKS® PROBLEMS

P1-1) Model the following object. Note that the dimensions are given in inches.

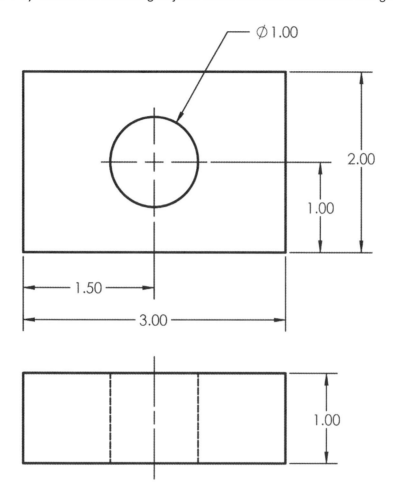

P1-2) Model the following object. Note that the dimensions are given in millimeters.

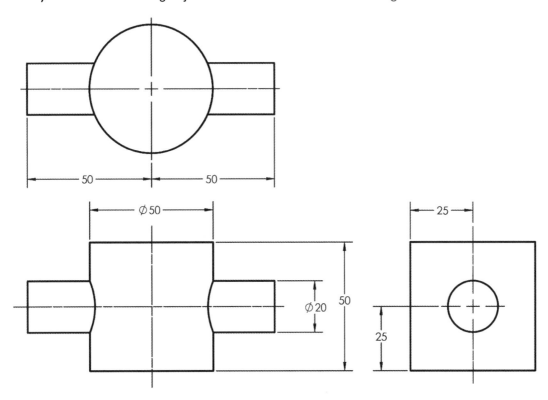

P1-3) Use SOLIDWORKS® to create a solid model of the following 1345 Aluminum part. Calculate the weight of the part. Dimensions are given in inches.

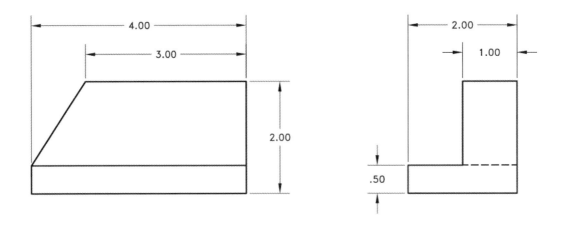

P1-4) Use SOLIDWORKS® to create a solid model of the following 6061 Aluminum part. Calculate the weight of your part. Dimensions are given in inches.

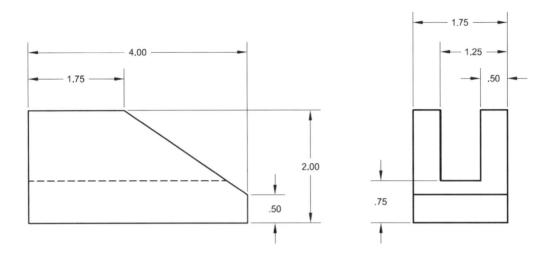

P1-5) Create a solid model of the following 1345 Aluminum part and calculate the weight of the part. Dimensions are given in inches.

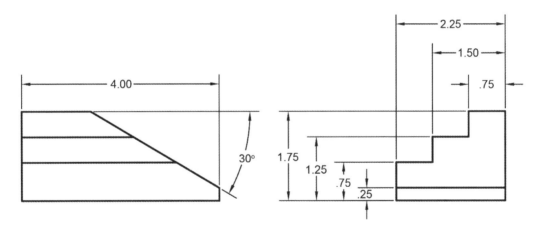

P1-6) Create a solid model of the following Brass part and calculate the mass of the part. Dimensions are given in millimeters.

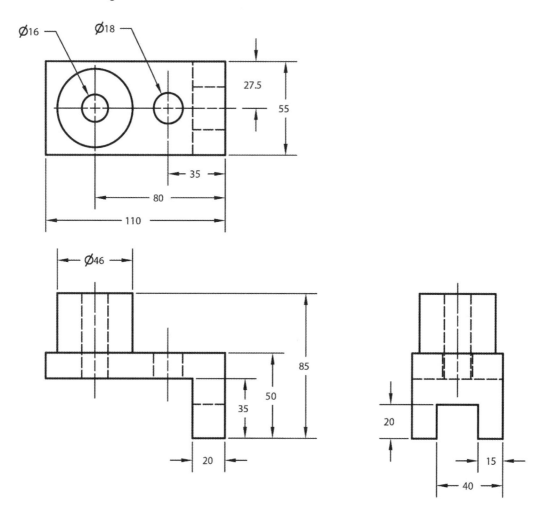

P1-7) Use SOLIDWORKS® to create a solid model of the following 1020 Steel part. Calculate the weight of the part. Dimensions given in inches.

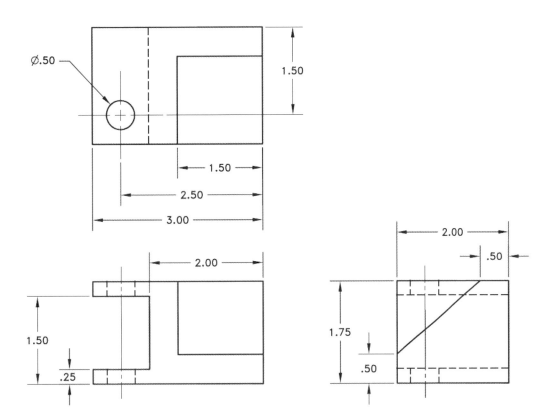

P1-8) Create a solid model of the following Gray Cast Iron part and calculate the weight of the part. Dimensions are given in inches.

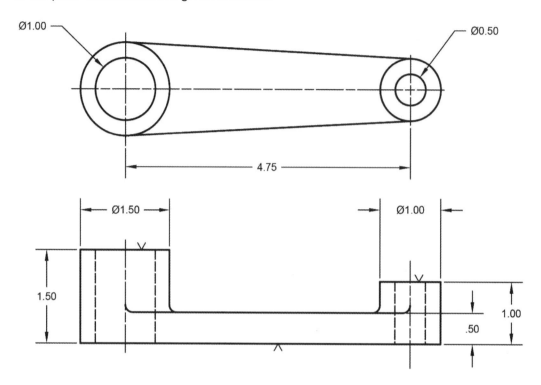

NOTE: ALL FILLETS AND ROUNDS R.12
UNLESS OTHERWISE SPECIFIED

P1-9) Use SOLIDWORKS® to create a solid model of the following 1020 Steel part. Calculate the weight of the part. Dimensions given in inches.

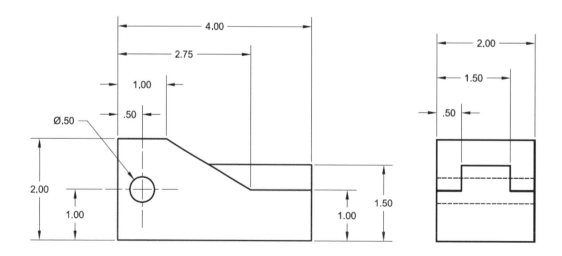

P1-10) Create a solid model of the following Oak part and calculate the weight of the part. Dimensions are given in inches.

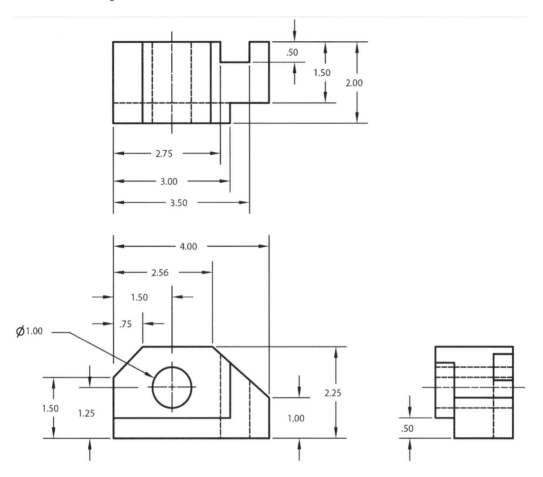

P1-11) Use SOLIDWORKS® to create a solid model of the following ABS plastic part. Calculate the weight of the part. Dimensions given in millimeters.

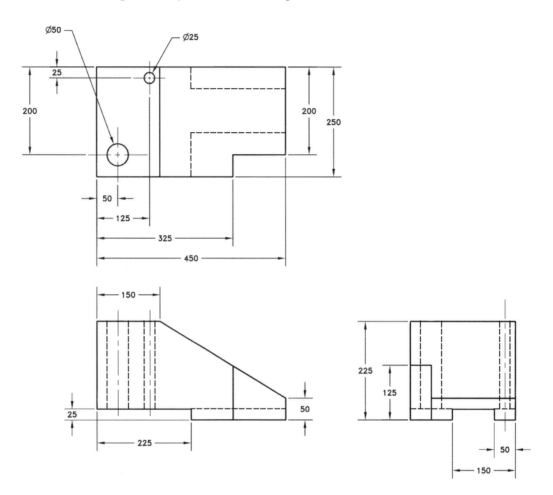

P1-12) Use SOLIDWORKS® to create a solid model of the following Oak part. Calculate the mass of the part. Dimensions given in millimeters.

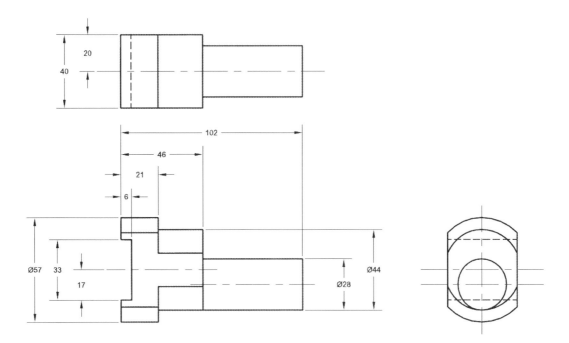

P1-13) Use SOLIDWORKS® to create a solid model of the following Grey Cast Iron. Calculate the mass of the part. Dimensions given in millimeters. Note that all fillets and rounds are R3.

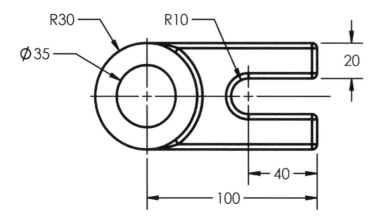

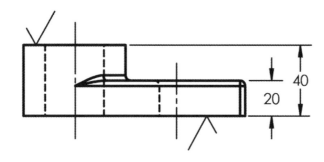

P1-14) Create a solid model of the following ABS plastic part and calculate the weight of the part. Dimensions are given in inches.

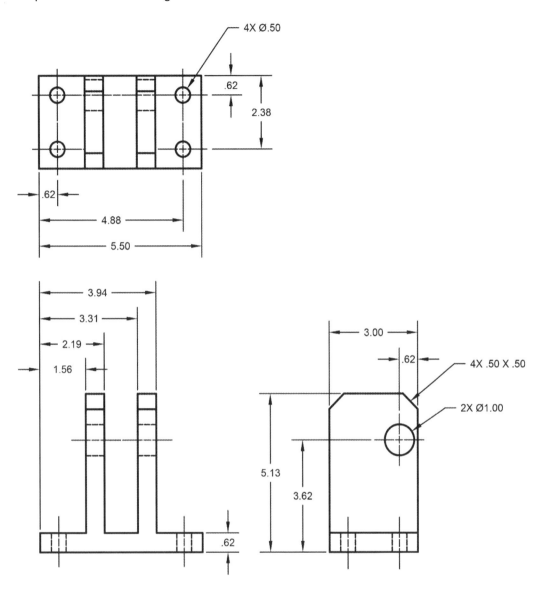

NOTES:

CHAPTER 2

BASIC DRAWINGS IN SOLIDWORKS®

CHAPTER OUTLINE

2.1) CREATING DRAWINGS IN SOLIDWORKS®..2
2.2) NEW DRAWING, USER INTERFACE, AND SETUP 2
    2.2.1) New drawing 2
    2.2.2) Selecting sheet size 2
    2.2.3) Model view 3
    2.2.4) Drawing user interface 5
2.3) DRAWING VIEWS...6
    2.3.1) Standard and projected views 6
2.4) ANNOTATIONS..7
    2.4.1) Centerlines and center marks 7
    2.4.2) Model Items 9
2.5) LAYERS 9
2.6) ANGLED BLOCK PRINT TUTORIAL...11
    2.6.1) Prerequisites 11
    2.6.2) What you will learn? 11
    2.6.3) Orthographic Projection 13
    2.6.4) Dimensions 16
    2.6.5) Filling in the title block..20
2.7) DEFINING A FRONT VIEW TUTORIAL 22
    2.7.1) Prerequisites 22
    2.7.2) What you will learn? 22
    2.7.3) The initial Orthographic Projection....................................23
    2.7.4) Changing the Front View 24
CREATING BASIC DRAWINGS IN SOLIDWORKS® PROBLEMS 27

CHAPTER SUMMARY

In this chapter you will learn how to create an orthographic projection in SOLIDWORKS®. You will also explore new feature-based commands such as Revolve and the Hole Wizard. By the end of this chapter, you will be able to create an orthographic projection of your part, show a pictorial on the print and edit the contents of the title block.

2.1) CREATING DRAWINGS IN SOLIDWORKS®

It is very simple to create an orthographic projection of a part in SOLIDWORKS®. Once a solid model is created, it just takes a few clicks to create an orthographic projection and pictorial of the model. One thing to keep in mind while using SOLIDWORKS® to create the part print is that the program does not always follow the ASME standard. So, it is your job to inspect the part print and make any adjustments to the drawing that is required for it to conform to the ASME standard.

2.2) NEW DRAWING, USER INTERFACE, AND SETUP

2.2.1) New drawing

To create a part print or orthographic projection of a part, first a model the object must be created. Then, to create a drawing of the part select the new document icon and then drawing . SOLIDWORKS® gives many options to choose from when creating a drawing. For example, sheet size, the number and the type of views can all be chosen. Also, the user can sketch on the drawing if content needs to be added.

2.2.2) Selecting sheet size

When creating a drawing, SOLIDWORKS® will first ask for a sheet size to be chosen. (This is the physical size of the paper where the drawing will be placed.) Each sheet size (e.g. A, B, C, D) has a different physical size. These sizes are controlled by the ASME Y14.110 standard for inch sizes and ASME Y14.110M for millimeter sizes. Figure 2.2-1 shows the *Sheet Format/Size* window and the different options available.

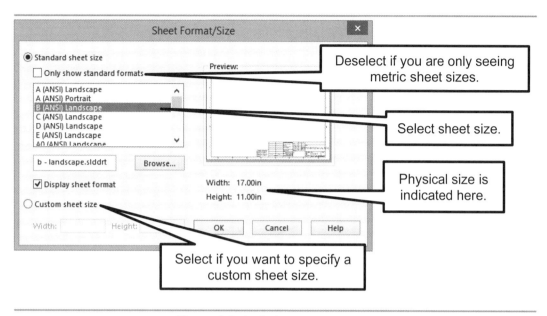

Figure 2.2-1: Selecting sheet size

2.2.3) Model view

A *Model View* window will appear which will allow the user to browse for a part or select an open part to create the drawing (see Figure 2.2-2). After the part is selected, SOLIDWORKS® will allow the selection of the number and type of views for display. The display style and scale of each view can also be chosen (see Figure 2.2-3). SOLIDWORKS® also gives the user the ability to create views after the drawing space has been entered by using the commands located in the ***View Layout*** tab.

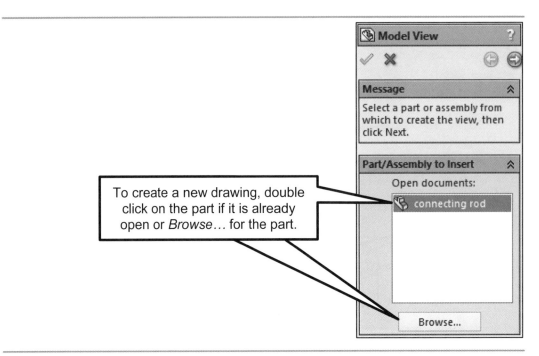

To create a new drawing, double click on the part if it is already open or *Browse…* for the part.

Figure 2.2-2: *Model View* Window

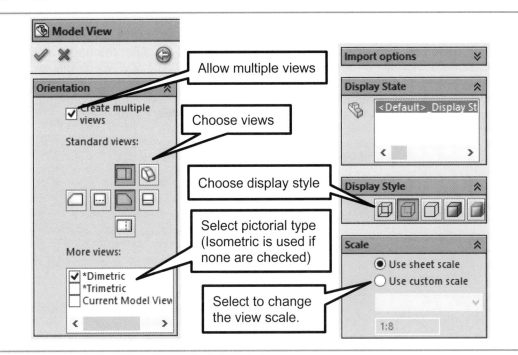

Figure 2.2-3: Model View Options

2.2.4) Drawing user interface

The drawing user interface has commands specific to creating a part print. The commands available allows the user to create views, sketch on them, and add dimensions and annotations. Figure 2.2-4 shows the drawing user interface. The different tabs available in the Command Manager are:

- **Drawing:** The *Drawing* tab contains frequently used commands.
- **View Layout:** The *View Layout* tab contains commands that gives the ability to add views to the drawing such as section views, detailed views and auxiliary views.
- **Annotation:** The *Annotation* tab contains commands that allow the views to be annotated, such as adding dimensions, notes, and GDT symbols.
- **Sketch:** The *Sketch* tab allows the user to sketch on and modify the drawing.
- **Evaluate:** The *Evaluate* tab contains commands that allow the user to check spelling, measure distances and mark up the drawing.
- **Sheet Format:** The *Sheet Format* tab contains commands that allow for editing of the sheet format, title block and border.

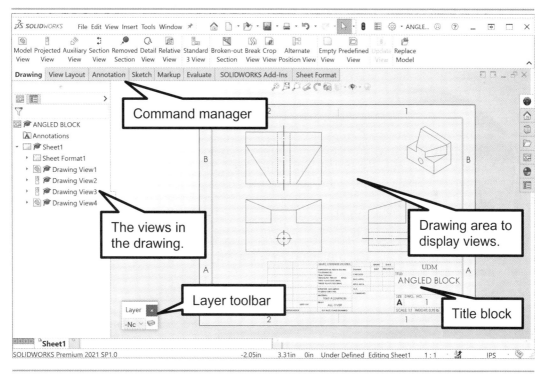

Figure 2.2-4: Drawing user interface

2.3) DRAWING VIEWS

When a new drawing is started, you select the model file that you wish to create the orthographic projection from and then you select the views you want to show. After you have entered the drawing or paper space, you can add additional views using the commands located in the *View Layout* tab (see Figure 2.5-1). Projected views, auxiliary views, section views, and detail views are some of the options available here.

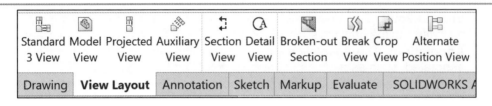

Figure 2.5-1: View Layout tab

2.3.1) Standard and projected views

If it is found that the user needs to start over with the views selected or just need to add an additional projected view, the **Standard 3 View** and **Projected View** commands can be used.

- **Standard 3 View** : This command creates the three standard orthographic views (i.e. front, top, right side). These views can be created using either third-angle or first-angle projection.

- **Projected View** : A projected view is an orthographic view created off of an existing view. The projected view can be created using either third-angle or first-angle projection.

2.4) ANNOTATIONS

Drawings may be annotated by using the commands located in the **Annotation** tab. Annotation commands include adding dimensions, notes, surface finish symbols, GDT notes and symbols, centerlines, and center marks. The *Annotation tab* is shown in Figure 2.4-1.

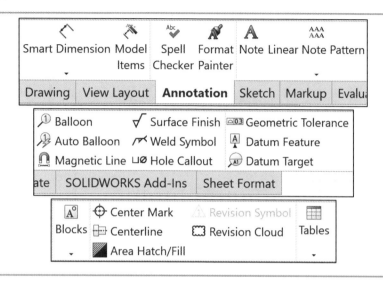

Figure 2.4-1: Annotation tab

2.4.1) Centerlines and center marks

When creating an orthographic projection, SOLIDWORKS® generally adds the center marks automatically, but does not add the centerlines. The user can add centerlines using the *Centerline* command. In many cases, even though SOLIDWORKS® has added center marks, these will need to be modified.

- **Centerlines** [⊞ Centerline] may be added automatically by selecting the **Select View** toggle and then selecting the view which will add the center lines to the entire view. They may also be added individually by selecting the two lines that define the centerline.

- **Center Marks** [⊕ Center Mark] may be added by selecting the circular feature where the center mark is wanted to be added. The *Center Mark* properties may be adjusted by changing the parameters in the *Center Mark* window (See Figure 2.3-1). You need to **deselect** the **Use document defaults** to do this. The length of the center mark may be adjusted by clicking on the center mark and dragging the grip boxes.

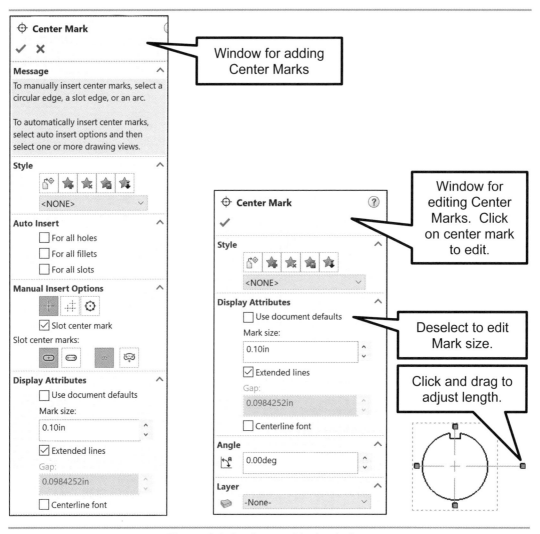

Figure 2.3-2: Center Mark window

2.4.2) Model Items

The *Model items* command allows the user to insert dimensions, annotations, and reference geometry from the model into the drawing. See Figure 2.3-3 to see all of the options available. After inserting model items, they can be deleted (**DELETE** key), dragged to another view (**SHIFT + drag**) or copied to another view (**CTRL + drag**). They also can be repositioned. For a smooth motion try using the **ALT** key while dragging. Each dimension also has grip boxes that will allow the ends of each extension line to be adjusted.

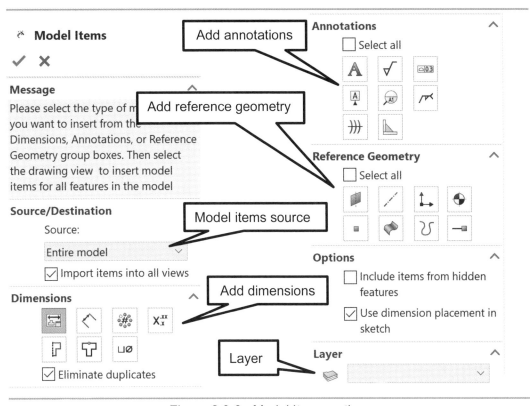

Figure 2.3-3: Model Items options

2.5) LAYERS

Layers are like transparencies, one placed over the top of another. Each transparency/layer may contain a different line type or a different part of the drawing. One layer may be used to create visible lines, while another layer may be used to create hidden lines. Assigning a different line type to each layer helps control and organize the drawing. If the Layer toolbar is not active, go to **View-Toolbars-Layer** to activate it. To create a new layer or edit a layer's properties, click on **Layer Properties** in the *Layer* toolbar. A *Layers* window will appear which will allow for the creation and editing of layers (see Figure 2.5-3).

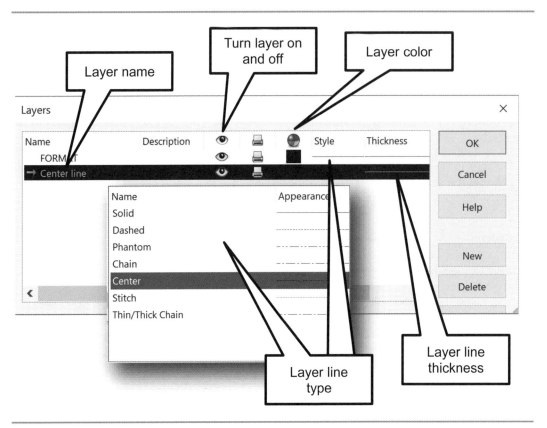

Figure 2.5-3: Layer Window

2.6) ANGLED BLOCK PRINT TUTORIAL

2.6.1) Prerequisites

To complete this tutorial, the user should have completed the listed tutorial and be familiar with the listed topics.

- Chapter 1 – Angled block tutorial
- Passing familiarity with orthographic projection.
- Ability to read dimensions.

2.6.2) What you will learn?

The objective of this tutorial is to introduce you to the SOLIDWORKS' drawing capabilities. In this tutorial, you will be creating a part print of the angled block that you modeled in Chapter 1. The part print is shown in Figure 2.6-1. Specifically, you will be learning the following commands and concepts.

Drawing

- New drawing
- Sheet size
- Standard 3 view
- Display style
- Center line
- Center mark
- Edit sheet format
- Edit sheet

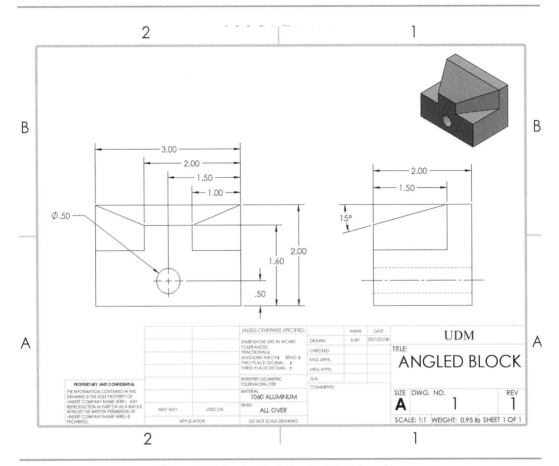

Figure 2.6-1: Angled block detail drawing

2.6.3) Orthographic Projection

It is often the case that CAD programs do not follow the ASME drawing standards when creating a part print. Therefore, it is important that the user reviews the print and decides what is correct and changes what needs to be adjusted.

1) **Open ANGLED BLOCK.SLDPRT**.

2) Start a **New Drawing** .

3) Select a **Sheet Size** of **A (ANSI) Landscape**. If only the metric sheet sizes can be seen, deselect the *Only show standard formats* check box.

> ➤ See section 2.2.2 to learn about **Selecting Sheet Sizes**.

4) Set the drafting standard to **ANSI** and set all the text to **upper case** for notes, tables, and dimensions. (*Options* ⚙ *– Document Properties – Drafting Standard*)

5) Set the units to **IPS** (i.e. inch, pound, second) and set the **Decimals = .12**. Also, select the rounding option, **Round half to even**. (*Options* ⚙ *– Document Properties – Units*)

6) **Save** the drawing as **ANGLED BLOCK.SLDDRW**. Remember to **Save** often throughout this tutorial.

7) In the **View Layout** tab, select **Model View**
. In the *Model View* window, double click on the
ANGLE BLOCK part and then choose the
following.

    a. Orientation: Select **Create multiple views** and
choose to create a **Front**, **Top, Right side** and
Isometric pictorial views.

    b. Display style = **Hidden Lines Visible**.

    c. Scale = **Use sheet scale**.

    d.

➤ See section 2.2.3 to learn about the ***Model view***
options.

---

8) Click and drag the views to the approximate
locations shown.

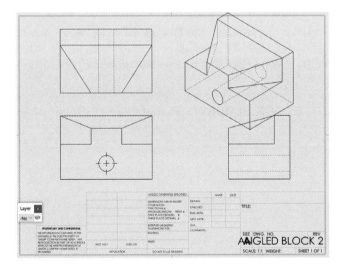

9) Notice that the isometric pictorial is very large. So, the size needs to be reduced. Select the pictorial view in the drawing area and set the display style as **Shaded with Edges** and use a **Use custom scale** and a **1:2** scale.

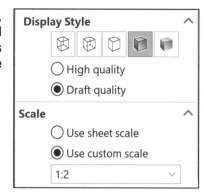

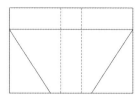

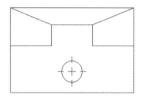

10) Add missing centerlines. Click on **Centerlines** `Centerline` in the *Annotation* tab. Select the **Select View** check box, and then select the **top view** and **right-side view**. Click **OK** in the *Centerline* window.

11) Adjust the center mark in the front view by clicking on it. Deselect the **Use document defaults** check box and adjust the *Mark size* to **0.07in**.

> ➤ See section 2.4.1 to learn about *Centerlines & Center Marks*.

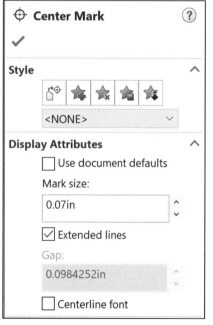

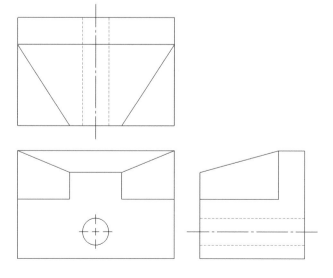

2.6.4) Dimensions

1) Add **Model Items** (*Annotation* tab) 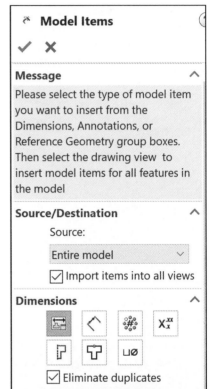 to the drawing using the following settings. Ignore the fact that the dimensions are jumbled and overlapping. They will be moved to more appropriate locations.
   a) Source = **Entire model**
   b) Activate the **Import items into all views** check box.
   c) Dimensions = **Marked for drawing dimensions**
   d) Activate the **Eliminate duplicates** check box.

> ➢ See section 2.4.2 to learn about the *Model items* options.
> ➢ Note: Don't worry, the dimensions may look messy. They may look similar to what is shown, or they may look entirely different. It is okay. This is where a good knowledge of Engineering Graphics comes in handy. Later, the print will be changed so that it is easily readable and presentable.

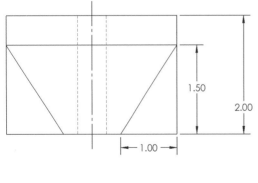

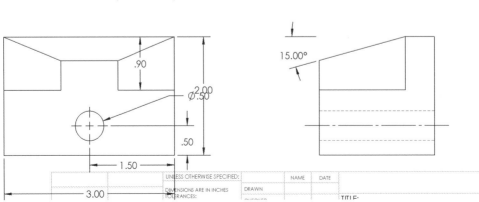

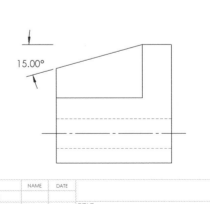

2) The *Front* view will be focused on first. Click and drag the dimensions to the approximate locations shown in the figure. If a dimension is not currently in the front view, click on the dimension, hold the **SHIFT** key down and drag it to the front view.

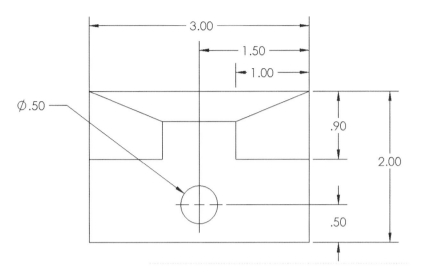

3) **Delete** the **.90** dimension and add the **1.10** and **2.00** dimensions using **Smart**

dimension .

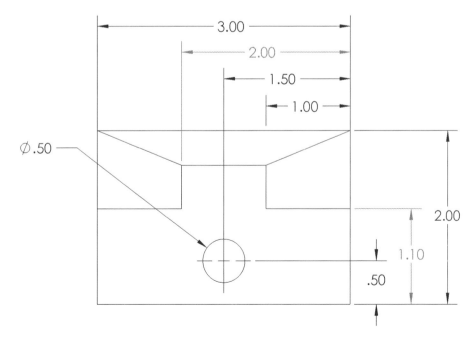

4) Notice that the added dimensions are grey. To make them black a layer must be created and the dimensions must be placed on it. If the *Layers* toolbar is not showing (it should be in the bottom left corner), activate it (**View – Toolbars – Layer**).

5) Create a **New** layer called **Dimension**.

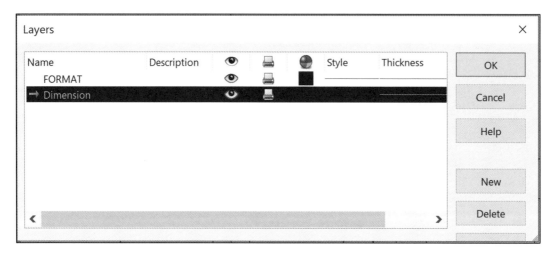

> ➤ See section 2.5 to learn about *Layers*.

6) Select the 2.00 and 1.10 dimension and then select the dimension layer in the Layer toolbar. This will place those two dimensions on that layer. The dimension should now be black.

7) Notice that the gaps between some of the extension lines and the object are not ideal. To change this, click on the dimension. Grip boxes will appear. Drag the grip boxes to an appropriate location. For examples, see the figure.

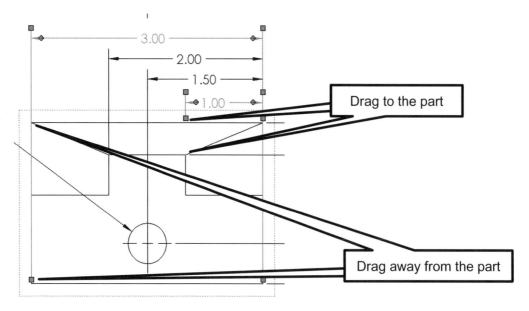

8) Drag the dimensions and the dimension text into the correct locations. Holding the **ALT** key will produce a smooth drag. The final *Front* view should look like the figure.

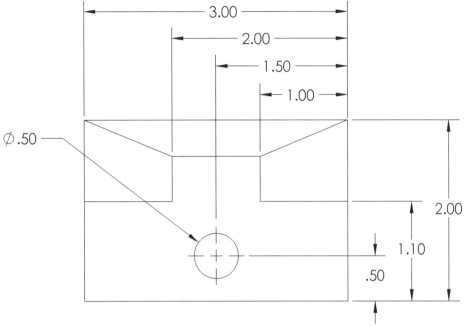

9) Using the **SHIFT** key, drag the dimensions in the *Top* view into the *Right-Side* view. Then, make any adjustments to make the Right-Side view look like the figure.

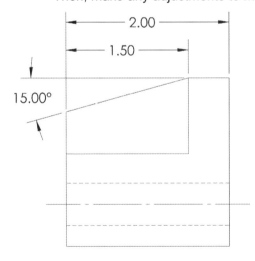

10) If the angle dimension has decimals, click on it and fix it in the *Dimension* options window.

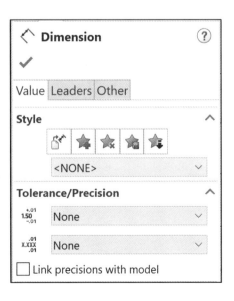

11) Notice that there are no dimensions on the *Top* view. We can keep it, but we would have to change the scale of the views. In this case, let's just delete the *Top* view. The pictorial will fill in any information lost by deleting the *Top* view.

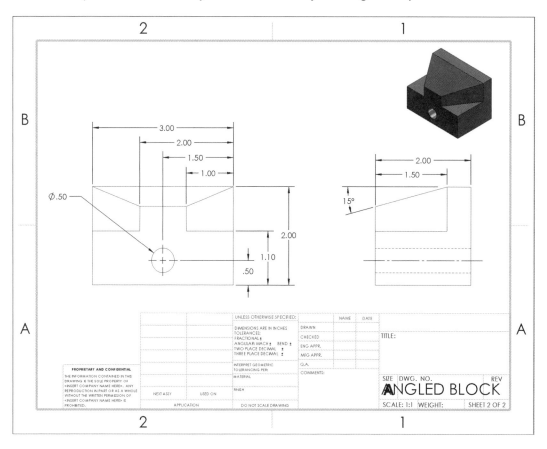

2.6.5) Filling in the title block

1) **Right-click** on the drawing somewhere outside the views but inside the drawing area and select **Edit Sheet Format**. Notice that the drawing will disappear.

2) Move the mouse around the title block and notice that every once in a while a text symbol will appear. This indicates a text field. To enter text, just double click on the field.

3) Enter the following text. Note that all text should be capitalized. You can use the formatting window to increase or decrease the text size as needed.
   a) TITLE = **ANGLE BLOCK**
   b) Above the title place your school's name.
   c) DWG. NO. = **1**
   d) REV = **1**
   e) WEIGHT = Enter the value listed in the *Mass Properties*
   f) DRAWN NAME = your initials
   g) DRAWN DATE = enter the date in this format (YYYY/MM/DD)
   h) MATERIAL = **1060 ALUMINUM**
   i) FINISH = **ALL OVER**

4) Get out of the title block by **right-clicking** on the drawing and selecting **Edit Sheet**.

5) **Save** your drawing.

2.7) DEFINING A FRONT VIEW TUTORIAL

2.7.1) Prerequisites

To complete this tutorial, the user should have completed the listed tutorial and be familiar with the listed topics.

- Chapter 2 – Angled block part print tutorial
- Passing familiarity with orthographic projection.

2.7.2) What you will learn?

The objective of this tutorial is to learn how to change the defined front view. The orthographic projection for a simple U-block is shown in Figure 2.7-1. The front view is not ideal. Specifically, the user will be learning how to change the view by learning the following commands and concepts.

Drawing

- Update Standard View

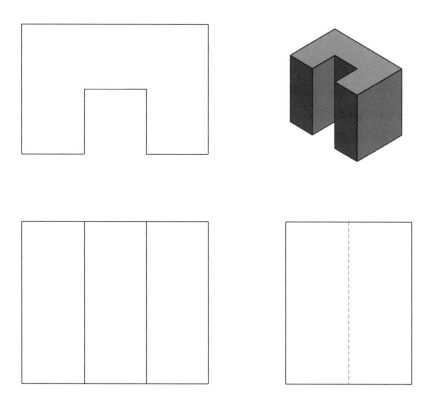

Figure 2.7-1: U-Block initial orthographic projection

2.7.3) The initial Orthographic Projection

1) **Open** the part file **U-BLOCK.SLDPRT** that came with this book.

2) Start a **New Drawing** .

3) Select a **Sheet Size** of **A4 (ANSI) Landscape**.

> ➢ See section 2.2.2 to learn about *Selecting Sheet Size*.

4) Double-click on **U-BLOCK** in the *Model View* window.

5) Create a **Front, Top,** and **Right-side view** along with an **Isometric pictorial** with a *Display style* of **Hidden Lines Visible**. Use the sheet scale.

6) Change the *Scale* of the pictorial to **1:2** and the *Display style* to **Shaded with Edges**.

7) Notice that the view that should be the Front view is currently the Top view. We need to change this.

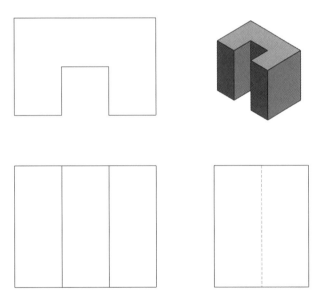

2.7.4) Changing the Front View

1) Enter the **U-BLOCK.SLDPRT** model file (**Window – U-BLOCK**).

2) Hit the **SPACE** bar to activate the *View Cube* and view the block from the view shown in the figure.

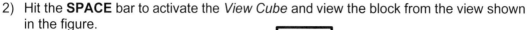

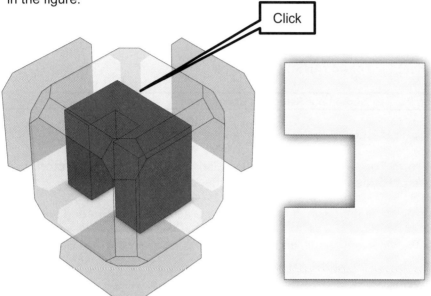

Click

3) Notice that the view is not in the correct orientation for the *Front* view. Hold down the **SHIFT** key and then hit the **Left arrow – Up arrow – Right arrow** in sequence. This should place the view in the proper orientation. If it does not, continue figuring out the proper arrow sequence until the *Front view* looks like the figure.

4) Hit the **SPACE BAR**. In the *Orientation* window, click on **Update Standard Views** and then click on **Front** view . Click **Yes** if a warning window appears.

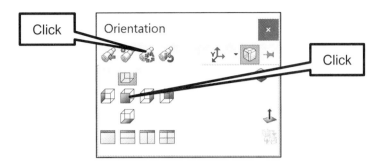

5) Enter the **U-BLOCK.SLDDRW** drawing file. Notice that the views have updated.

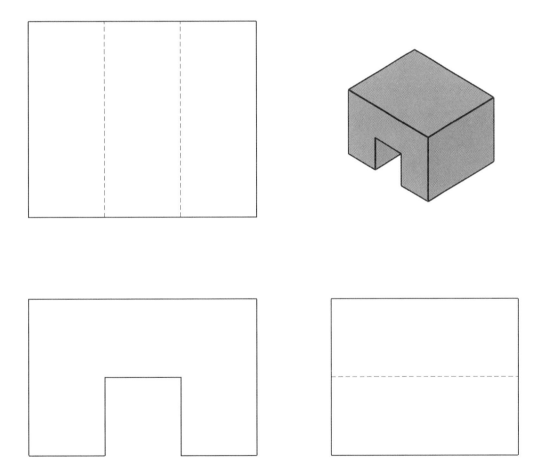

NOTES:

CREATING BASIC DRAWINGS IN SOLIDWORKS® PROBLEMS

P2-1) Use SOLIDWORKS® to create the part print that looks exactly like the detailed drawing of the model shown in P1-1. Fill in the appropriate information into your title block. Enter the weight of the part into the title block of the drawing.

P2-2) Use SOLIDWORKS® to create the part print that looks exactly like the detailed drawing of the model shown in P1-2. Fill in the appropriate information into your title block. Enter the weight of the part into the title block of the drawing.

P2-3) Use SOLIDWORKS® to create the part print that looks exactly like the detailed drawing of the model shown in P1-3. Fill in the appropriate information into your title block. Enter the weight of the part into the title block of the drawing.

P2-4) Use SOLIDWORKS® to create the part print that looks exactly like the detailed drawing of the model shown in P1-4. Fill in the appropriate information into your title block. Enter the weight of the part into the title block of the drawing.

P2-5) Use SOLIDWORKS® to create the part print that looks exactly like the detailed drawing of the model shown in P1-5. Fill in the appropriate information into your title block. Enter the weight of the part into the title block of the drawing.

P2-6) Use SOLIDWORKS® to create the part print that looks exactly like the detailed drawing of the model shown in P1-6. Fill in the appropriate information into your title block. Enter the weight of the part into the title block of the drawing.

P2-7) Use SOLIDWORKS® to create the part print that looks exactly like the detailed drawing of the model shown in P1-7. Fill in the appropriate information into your title block. Enter the weight of the part into the title block of the drawing.

P2-8) Use SOLIDWORKS® to create the part print that looks exactly like the detailed drawing of the model shown in P1-8. Fill in the appropriate information into your title block. Enter the weight of the part into the title block of the drawing.

P2-9) Use SOLIDWORKS® to create the part print that looks exactly like the detailed drawing of the model shown in P1-9. Fill in the appropriate information into your title block. Enter the weight of the part into the title block of the drawing.

P2-10) Use SOLIDWORKS® to create the part print that looks exactly like the detailed drawing of the model shown in P1-10. Fill in the appropriate information into your title block. Enter the weight of the part into the title block of the drawing.

P2-11) Use SOLIDWORKS® to create the part print that looks exactly like the detailed drawing of the model shown in P1-11. Fill in the appropriate information into your title block. Enter the weight of the part into the title block of the drawing.

P2-12) Use SOLIDWORKS® to create the part print that looks exactly like the detailed drawing of the model shown in P1-12. Fill in the appropriate information into your title block. Enter the weight of the part into the title block of the drawing.

NOTES:

CHAPTER 3

INTERMEDIATE PART MODELING IN SOLIDWORKS®

CHAPTER OUTLINE

3.1) SKETCH 2

    3.1.1) Trim Entities 2

    3.1.2) Convert Entities & Intersection Curve 4

    3.1.3) Offset Entities 6

    3.1.4) Construction Geometry 7

    3.1.5) Dimensioning using a centerline 8

3.2) FEATURES 9

    3.2.1) Revolve and Revolve cut 10

    3.2.2) Holes 12

    3.2.3) Patterns 14

    3.2.4) Mirror 15

    3.2.5) Reference Geometry 17

3.3) FLANGED COUPLING TUTORIAL 18

    3.3.1) Prerequisites 18

    3.3.2) What you will learn? 18

    3.3.3) Setting up the project 20

    3.3.4) Base revolve 20

    3.3.5) Adding features 24

3.4) TABLETOP MIRROR TUTORIAL 27

    3.4.1) Prerequisites 27

    3.4.2) What you will learn? 27

    3.4.3) Setting up the Mirror Face 29

    3.4.4) Modeling the Mirror Face 29

    3.4.5) Setting up the Mirror Base 40

    3.4.6) Modeling the Mirror Base 40

INTERMEDIATE PART MODELING IN SOLIDWORKS® PROBLEMS 49

CHAPTER SUMMARY

In this chapter you will learn how to create intermediate models in SOLIDWORKS®. You will explore new sketch-based commands such as Convert Entities and Trim and new feature-based commands such as Revolve and the Hole Wizard. By the end of this chapter, you will be able to create more realistic parts.

3.1) SKETCH

In Chapter 1 we discussed that before you can create a solid, you need to create a two-dimensional sketch that will define the solid's shape. Many different sketch elements are available in the **Sketch** tab shown in Figure 3.1-1. In this chapter, we will learn how to use the **Trim Entities**, **Convert Entities**, and **Offset Entities** commands.

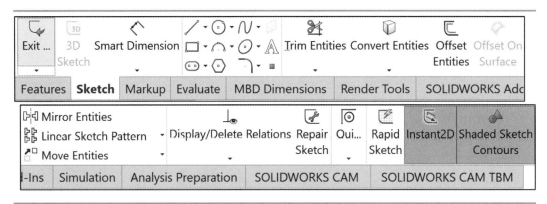

Figure 3.1-1: Sketch tab

3.1.1) Trim Entities

The **Trim Entities** command allows you to delete portions of your sketch that exist outside a boundary or between two elements. There are several different ways to trim based on your needs. I find that *Trim to closest* works for all common trim situations. Figures 3.1-2 through 3.1-4 show some of the trim effects. Listed are some other types of trims and what they do.

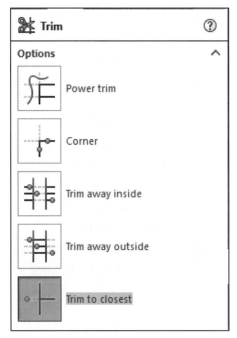

- Trim to closest: Trims or extends an entity up to the closest intersection.
- Power Trim
  - Extends sketch entities.
  - Trim single sketch entities to the nearest intersecting entity as you drag the pointer.
  - Trim one or more sketch entities to the nearest intersecting entity as you drag the pointer and cross the entity.

<u>Corner:</u> Modifies two selected entities until they intersect at a virtual corner.

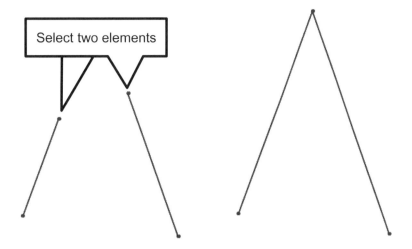

Figure 3.1-2: Corner

<u>Trim away inside:</u> Trims an entity between two intersections.

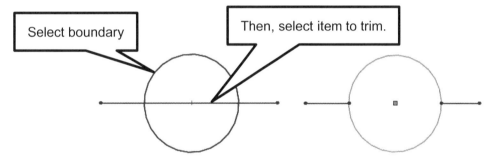

Figure 3.1-3: Trim away inside

Trim away outside: Trims an entity outside of two intersections.

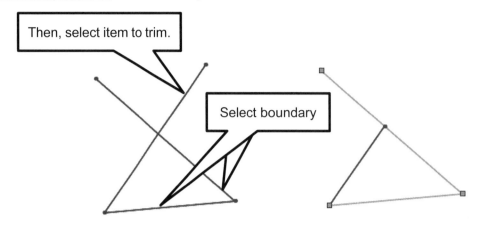

Figure 3.1-4: Trim away outside

3.1.2) Convert Entities & Intersection Curve

The **Convert Entities** [Convert Entities] and **Intersection Curve** [Intersection Curve] commands allow you to create sketches based on existing geometry. The *Convert Entities* command converts solid geometry edges or sketch entities to a sketch on a selected plane. The *Intersection Curve* command creates a sketch of the intersection between planes, solid bodies, and surface bodies. Figure 3.1-5 shows an example of the use of the *Convert Entities* command and Figure 3.1-6 shows an example of the use of the *Intersection Curve* command. To apply the *Convert Entities* and *Intersection Curve* commands use the following steps.

Convert Entities

1) Select a plane or face to sketch on.
2) Select the **Convert Entities** [Convert Entities] command.
3) Select the faces and/or edges you wish to convert.

Intersection Curve

1) Select at least two entities such as planes, faces, or surfaces. If you are sketching on a plane, the plane is automatically selected as one of the items.
2) Select the **Intersection Curve** [Intersection Curve] command.

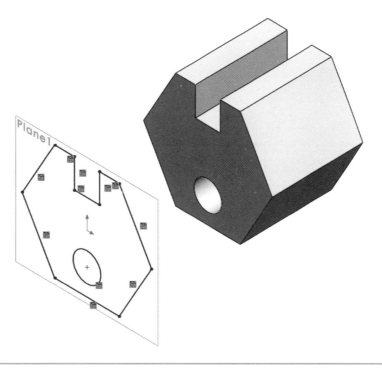

Figure 3.1-4: Convert Entities

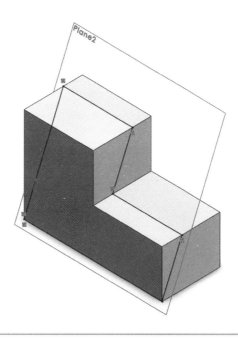

Figure 3.1-5: Intersection Curve

3.1.3) Offset Entities

The Offset Entities command allows you to offset a sketch entity, model face, or model edge a specified distance. See Figures 3.1-6 and 3.1-7 to learn how to offset a model face and edge. A sketch entity can be offset in a very similar way.

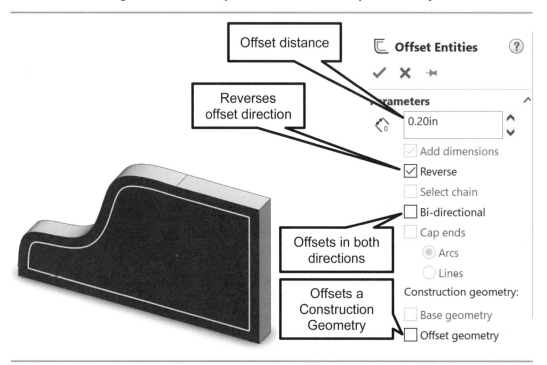

Figure 3.1-6: Offsetting a Model Face

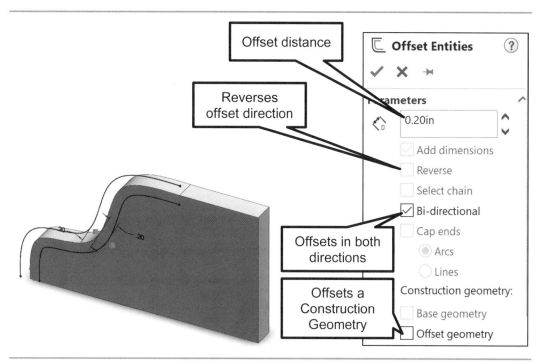

Figure 3.1-7: Offsetting a Model Edge

3.1.4) Construction Geometry

Construction geometries are used as construction aids. They are not used in the creation of the solid features. They just assist you in creating the entities that will be used to create the solid features. Construction geometries are shown as center lines. Construction geometries are also created when drawing certain sketch entities such as a *Center Rectangle* □ Center Rectangle as shown in Figure 3.1-8. You can also change standard geometry into construction geometry. Notice that the right side of the center rectangle shown in Figure 3.1-8 has been changed to a construction geometry. The steps used to change a standard entity to a construction geometry are as follows.

- **Construction Geometry:** ⌶⇄ Changes a regular sketch into a construction geometry sketch and vice versa. To change a sketch, right-click on the sketch and select the *Construction Geometry* command.
- **Centerline:** ⋮ Centerline Centerlines are automatically created as construction geometry.

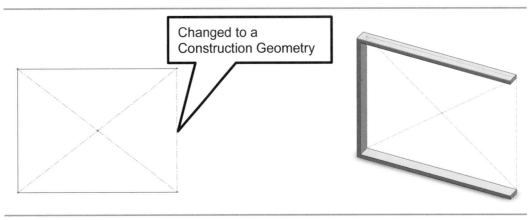

Changed to a Construction Geometry

Figure 3.1-8: Center Rectangle

3.1.5) Dimensioning using a centerline

If you are using a centerline as a revolve axis, you may want to dimension diameters instead of radii. For example, Figure 3.1-9 shows a revolved part dimensioned using radius dimensions and dimensioned using diameter dimensions. To dimension using diameter dimensions follow these steps.

Dimensioning using a centerline

1) Select the **Smart Dimension** command.
2) Select the centerline and then the feature you wish to dimension.
3) Pull the dimension to the opposite side of the centerline and click your left mouse button. The dimension should automatically change to a diameter dimension.

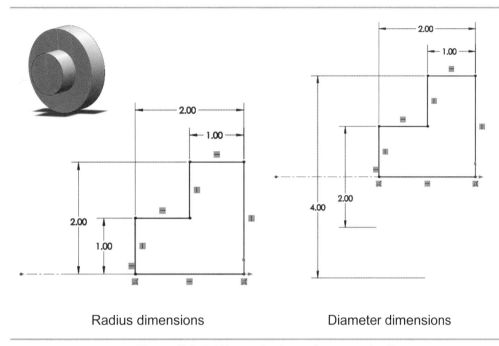

Radius dimensions   Diameter dimensions

Figure 3.1-9: Dimensioning using a center line

3.2) FEATURES

In Chapter 1 the feature-based commands of **Extrude**, **Extrude Cut**, **Fillet,** and *Chamfer* were introduced. In this chapter, the **Revolve**, **Mirror**, **Hole**, and **Pattern** commands will be introduced. **Reference Geometries** will also be used. The feature commands are located in the **Feature** tab shown in Figure 3.2-1.

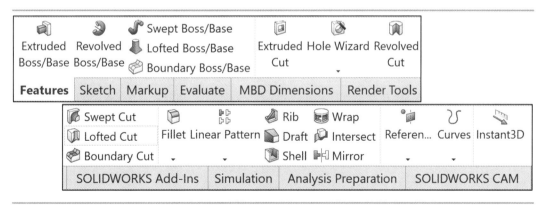

Figure 3.2-1: Feature tab

3.2.1) Revolve and Revolve cut

The **Revolve Boss/Base** [Revolved Boss/Base] and **Revolve Cut** [Revolved Cut] commands are located in the *Features* tab. The revolve commands take a closed profile and revolve it about a specified axis. The *Revolve Boss/Base* and the *Revolve Cut* commands work essentially the same way. The *Revolve Boss/Base* adds material and the *Revolve Cut* removes material. Note that the command **View – Hide/Show - Temporary Axes** may be used to show existing axes that may then be used as the revolve axis. The *Revolve* and *Revolve Cut* options are shown in Figure 3.2-2. An open profile (i.e. a profile that does not start and end in the same location) produces a thin feature revolve. The thin revolve options are shown in Figure 3.2-3.

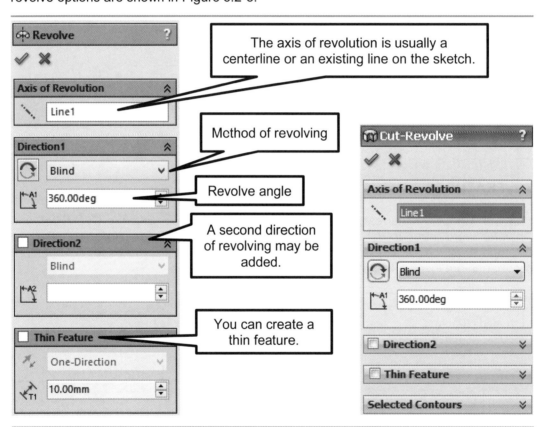

Figure 3.2-2: Revolve and Revolve cut options

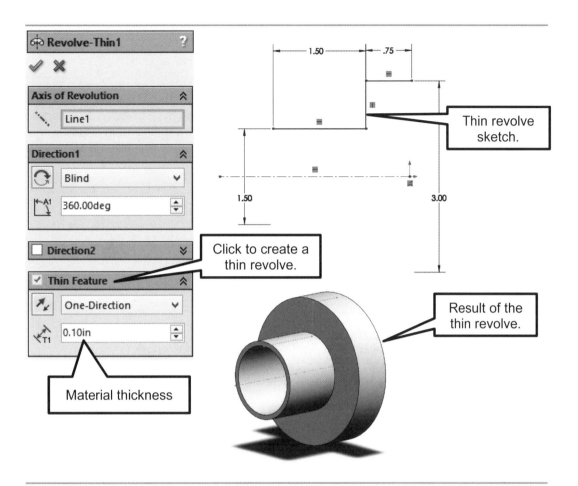

Figure 3.2-3: Thin Revolve options

3.2.2) Holes

The **Hole Wizard** command is in the *Features* tab. The *Holes* command allows you to create various types of holes in the part. The types of holes that can be created are shown in Figure 3.2-4 and the holes option window is shown in Figure 3.2-5. Each type of hole may be associated with a standard fastener or drill size. To create a *Hole*, use the following steps.

Creating Holes

1) Select the **Hole Wizard** command.
2) Select the hole shape by selecting one of the pictures (e.g., counterbore, countersink, blind). *See the option window on the next page.*
3) Select the standard (e.g., ANSI Metric, ANSI inch).
4) Select the hole type (e.g., screw clearance, tap drill, letter drill size).
5) Select the hole specification (e.g., drill size, associated fastener).
6) Specify the end condition.
7) Click on the **Positions** tab and then select the surface on which the hole will be placed.
8) Use **Dimensions** and **Sketch Relations** to position the hole.

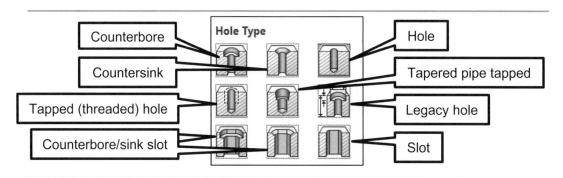

Figure 3.2-4: Types of holes that can be created.

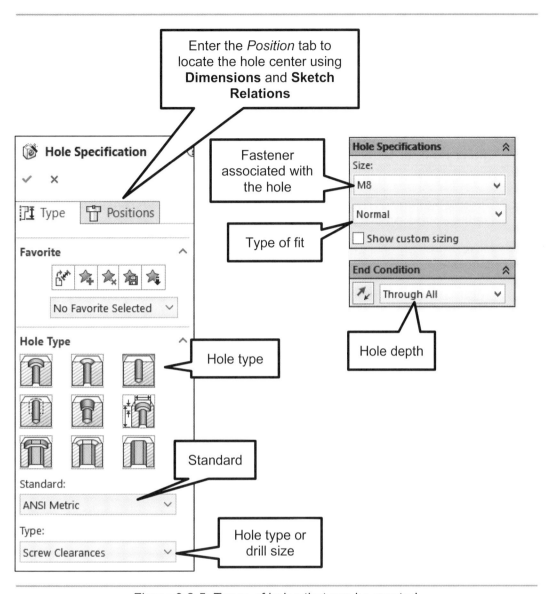

Figure 3.2-5: Types of holes that can be created.

3.2.3) Patterns

The **Pattern** commands are in the *Features* tab. They allow the user to array an existing feature. The most used patterns are *Linear Pattern* and *Circular Pattern*. However, there are other methods of patterning a feature as listed below. Figure 3.2-6 shows an example of a linear and circular pattern. The options available in the linear and circular pattern commands are shown in Figure 3.2-7.

- **Linear Pattern:** Linear Pattern The *Linear Pattern* command will repeat a feature along two directions.
- **Circular Pattern:** Circular Pattern The *Circular Pattern* command repeats a feature along a circular path.
- **Mirror:** Mirror The *Mirror* command creates a duplicate feature about a mirror line.
- **Curve Driven Pattern:** Curve Driven Pattern The *Curve Driven Pattern* command allows you to array along a predefined curve.
- **Sketch Driven Pattern:** Sketch Driven Pattern The *Sketch Driven Pattern* command uses sketch points. The feature is copied to each point in the sketch.
- **Table Driven Pattern:** Table Driven Pattern The *Table-Driven Pattern* command uses X-Y coordinates, to specify the pattern locations.
- **Fill Pattern:** Fill Pattern The *Fill Pattern* command fills a selected area with the patterned feature.
- **Variable Pattern:** Variable Pattern The *Variable Pattern* command allows the user to create patterns that have varying dimensions.

Circular Pattern   Linear Pattern

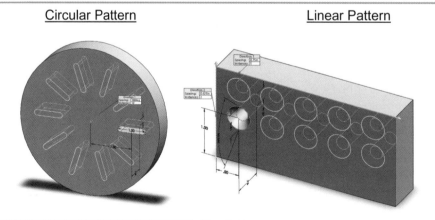

Figure 3.2-6: Linear and circular pattern examples

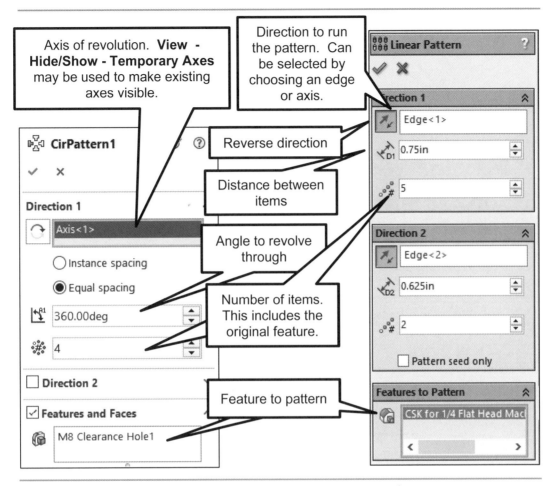

Figure 3.2-7: Linear and circular pattern options

3.2.4) Mirror

The **Mirror** [Mirror] command allows the user to duplicate a solid with its mirror image about a selected plane or face. An example of a mirrored object is shown in Figure 3.2-8. The Mirror option window is shown in Figure 3.2-9. Use the following steps to mirror an object.

Using the *Mirror* command

1) Select the **Mirror** [Mirror] command.
2) Select the mirror face or plane.
3) Select the features and/or faces to mirror.

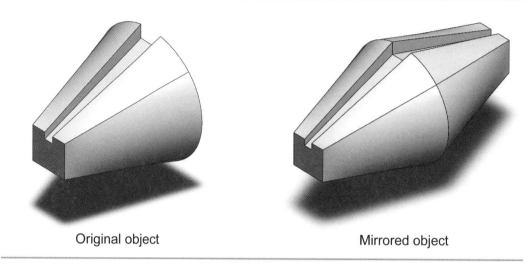

Original object   Mirrored object

Figure 3.2-8: Mirror example

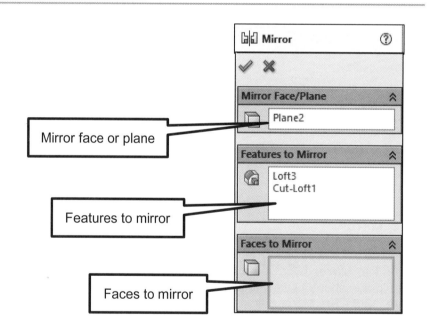

Mirror face or plane

Features to mirror

Faces to mirror

Figure 3.2-9: Linear and circular pattern examples

3.2.5) Reference Geometry

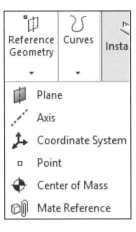

Reference Geometries are used as drawing aids. They are not directly used to create solid geometry. The reference geometries that are available in SOLIDWORKS® are *Plane*, *Axis*, *Coordinate System*, *Point*, *Center of Mass*, and *Mate Reference*.

There are many ways to define location and orientation of reference geometries. There are too many options to cover them all here. The best way to learn the different methods is to practice all of the options. If there is a particular orientation that is difficult to create, go to the SOLIDWORKS® help forum, the videos that come with this book, or to YouTube®.

3.3) FLANGED COUPLING TUTORIAL

3.3.1) Prerequisites

Before starting this tutorial, the following tutorials should already have been completed.

- Chapter 1 – Angled block tutorial
- Chapter 2 – Connecting Rod tutorial

It will also help if the user already has the following knowledge.

- A familiarity with threads and fasteners.

3.3.2) What you will learn?

The objective of this tutorial is to introduce the user to the **revolve** command, simple **patterns**, **holes**, and **section views**. In this tutorial, the *left coupling* for the *Flanged Coupling* assembly shown in Figure 3.3-1 will be modeled and the following commands and concepts will be learned.

Sketching

- Dimensioning using a centerline

Features

- Revolve
- Hole wizard
- Patterns

View

- ➤ Rotate

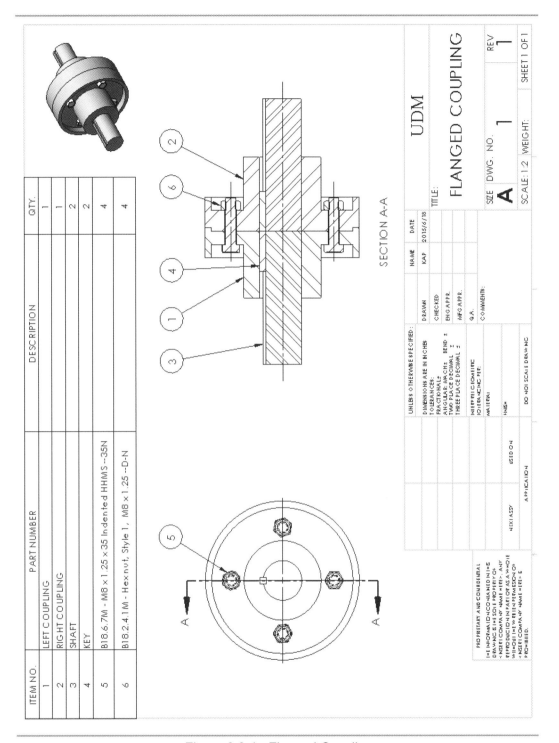

ITEM NO.	PART NUMBER	DESCRIPTION	QTY.
1	LEFT COUPLING		1
2	RIGHT COUPLING		1
3	SHAFT		2
4	KEY		2
5	B18.6.7M - M8 × 1.25 × 35 Indented HHMS --35N		4
6	B18.2.4.1M - Hexnut, Style 1, M8 × 1.25 --D-N		4

SECTION A-A

	NAME	DATE
DRAWN	KAP	2015/6/18
CHECKED		
ENG APPR.		
MFG APPR.		
Q.A.		
COMMENTS:		

UNLESS OTHERWISE SPECIFIED:

DIMENSIONS ARE IN INCHES
TOLERANCES:
FRACTIONAL±
ANGULAR: MACH±   BEND ±
TWO PLACE DECIMAL   ±
THREE PLACE DECIMAL ±

INTERPRET GEOMETRIC
TOLERANCING PER:

MATERIAL

FINISH

DO NOT SCALE DRAWING

PROPRIETARY AND CONFIDENTIAL

THE INFORMATION CONTAINED IN THIS
DRAWING IS THE SOLE PROPERTY OF
<INSERT COMPANY NAME HERE>. ANY
REPRODUCTION IN PART OR AS A WHOLE
WITHOUT THE WRITTEN PERMISSION OF
<INSERT COMPANY NAME HERE> IS
PROHIBITED.

NEXT ASSY   USED ON

APPLICATION

UDM

TITLE:

FLANGED COUPLING

SIZE DWG. NO.   REV
A   1   1

SCALE: 1:2 WEIGHT: SHEET 1 OF 1

Figure 3.3-1: Flanged Coupling

3.3.3) Setting up the project

1) **Start SOLIDWORKS** and start a **new part** .

2) Set the drafting standard to **ANSI** and set all of the text to **upper case** for notes, tables, and dimensions. (*Options* ⚙ – *Document Properties* – *Drafting Standard*)

3) Set the units to **MMGS** (millimeters, grams, second) and the **decimal = 0.1.** (*Options* ⚙ – *Document Properties* – *Units*), Set Decimal Rounding to: **Round half to Even.**

4) Save the part as **COUPLING.SLDPRT** (**File – Save**). Remember to save often throughout this project.

3.3.4) Base revolve

1) **Sketch** 🔲 *Sketch* on the **Right Plane**.

2) Draw a horizontal **Centerline** ✏️ *Centerline* starting at the origin and going off to the left. The *Centerline* command is located underneath the *Line* command.

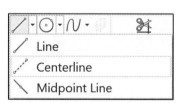

3) Use **Lines** 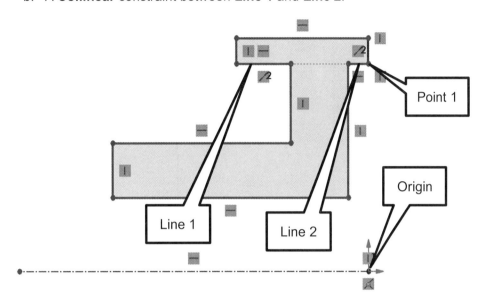 to create the following sketch.

4) Apply the following **Sketch Relations**.
    a. A **Vertical** constraint between the **Origin** and **Point 1**.
    b. A **Collinear** constraint between **Line 1** and **Line 2**.

5) **Dimension** [Smart Dimension] the sketch. To dimension using the centerline, select the centerline and then the feature, pull the dimension below the centerline to show the diameter. After you dimension the sketch, all lines should be black.

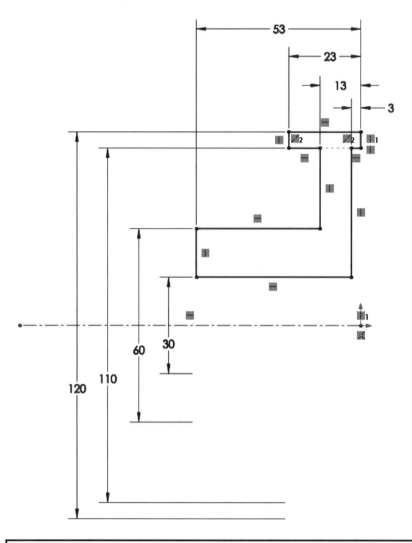

> ➢ See section 3.1.5 to learn how to *Dimension using a centerline*.

6) **Revolve Boss/Base** 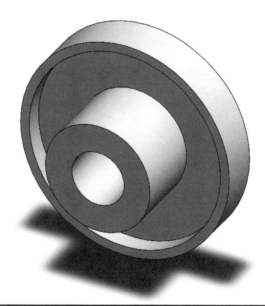 the sketch **360** degrees using the centerline as the axis of revolution and the **Blind** method.

> ➤ See 3.2.1 to learn how to use the ***Revolve Boss/Base & Revolve Cut*** command.

7) Rotate the part to look at the back side. Use the view cube (**Space bar**) or click and hold the middle button on the mouse and move the mouse.

8) Return to the previous view showing the front side using the view cube (**Space bar**) or **Ctrl + 7**.

9) **Save** your part.

3.3.5) Adding features

1) **Sketch** [Sketch] and **Dimension** [Smart Dimension] the square shown, on the front face of the object. Remember, **Ctrl + 8** gives the normal or straight on view.

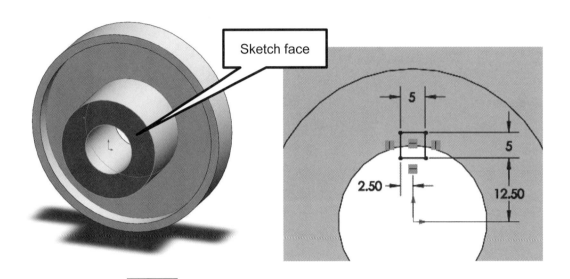

Sketch face

2) **Extrude Cut** [Extruded Cut] the sketch **Through All**.

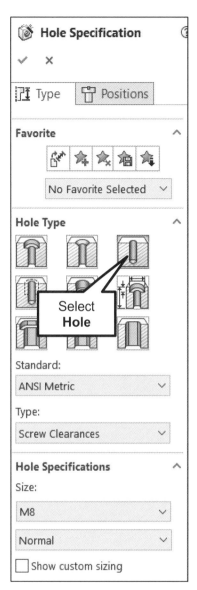

3) Add a normal **Clearance Hole** for a **M8** machine screw in the position shown. Make the origin of the hole **Vertical** with the part origin.
   a. Select the **Hole Wizard** command.
   b. Hole Type = **Hole**
   c. Standard = **ANSI Metric**
   d. Type = **Screw Clearances**
   e. Size: = **M8** and **Normal**
   f. Select the *Position* tab.
   g. Select the face where the hole will be located.
   h. Place and dimension the location of the hole center.

> ➢ See section 3.2.2 to learn how to create *Holes*.

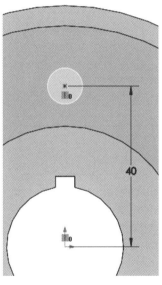

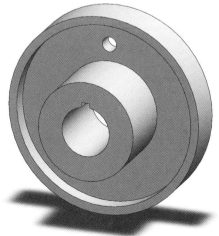

4) Create a **Circular Pattern** 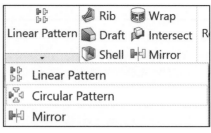 patterning the **Hole** that was just created to create a total of 4 holes. The *Circular Pattern* command is located under the *Linear Pattern* command. Use the center axis of the part as the revolving axis. Revolve **4** holes using a total angle of **360** degrees. To view the center axis, select **View – Hide/Show - Temporary Axes**.

➢ See section 3.2.3 to learn how to create *Patterns*.

5) Make the part out of **AISI 1020 Steel**.

6) **IMPORTANT! Save** this part and keep it. It will be used in future tutorials.

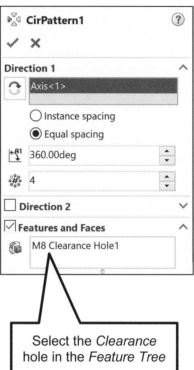

Select the *Clearance* hole in the *Feature Tree*

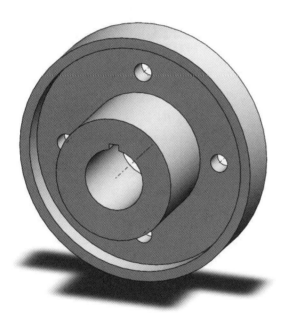

3.4) TABLETOP MIRROR TUTORIAL

3.4.1) Prerequisites

Before starting this tutorial, the following tutorials should already have been completed.

- Chap 3 – Flanged Coupling tutorial

3.4.2) What you will learn?

The objective of this tutorial is to introduce the **Convert Entities**, **Trim**, **Offset**, and **Mirror** commands.   **Reference Planes** and **Construction Geometry** will also be introduced. The *Tabletop Mirror* shown in Figure 3.4-1 will be modeled and the following commands and concepts will be covered.

Sketching

- Convert Entities
- Trim
- Offset Entities
- Construction Geometry

Features

- Revolve
- Mirror
- Reference Planes

Figure 3.4-1: Tabletop Mirror

3.4.3) Setting up the Mirror Face

1) **Start SOLIDWORKS** and start a **new part** .

2) Set the drafting standard to **ANSI** and set the text to **upper case** for notes, tables, and dimensions. (*Options* ⚙ – *Document Properties – Drafting Standard*)

3) Set the units to **IPS** (inch, pound, second) and the **decimal = 0.123.** (*Options* ⚙ – *Document Properties – Units*)

4) Save the part as **MIRROR FACE.SLDPRT** (**File – Save**). Remember to save often throughout this tutorial.

3.4.4) Modeling the Mirror Face

1) **Sketch** ⌐ Sketch on the **Front Plane**.

2) Draw 2 concentric **Circles** Ⓞ Circle starting at the origin and **Dimension** the sketch.

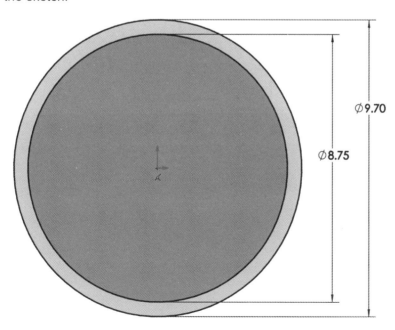

Ø9.70

Ø8.75

3) **Extrude**  the sketch **0.375** inches.

4) **Sketch** on the back face of the Extrude and use the **Convert Entities**

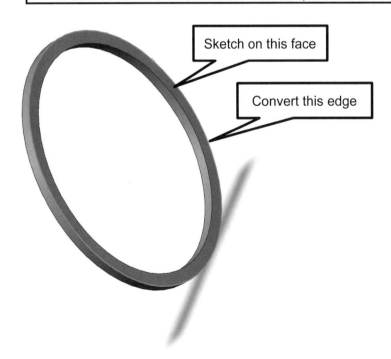
command to make the outer circle a sketch. To convert the outer circle to a sketch, select the outer circle and then select the *Convert Entities* command.

> ➤ To learn about the ***Convert Entities*** command, see section 3.1.2.

Sketch on this face

Convert this edge

5) **Extrude**  the sketch **0.125** inches. By default, this extrude should travel away from the original extrude. This is what we want.

Boss-Extrude2

From
Sketch Plane

Direction 1
Blind

0.125in

☑ Merge result

6) **Sketch** 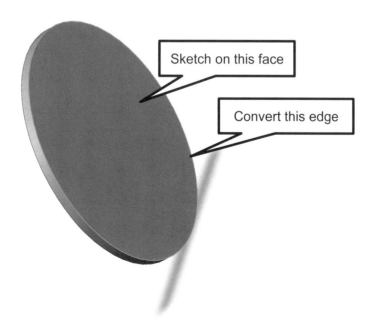 on the back face of the Extrude and use the **Convert Entities** command to make the outer circle a sketch.

Convert Entities

Sketch on this face

Convert this edge

7) Without leaving the sketch, draw a **Circle** [Circle] starting at the origin and **Dimension** [Smart Dimension] the sketch. Then, draw a horizontal **Line** [Line] passing through the origin as shown.

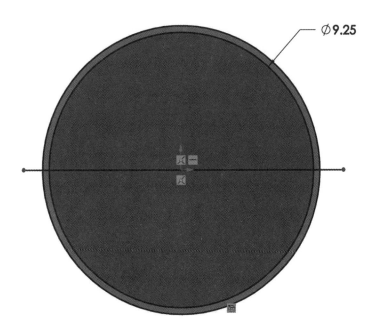

⌀**9.25**

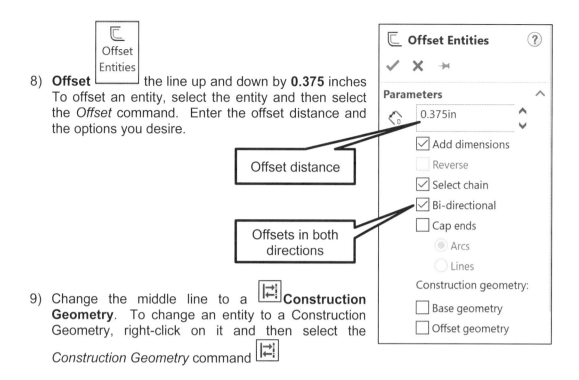

8) **Offset** the line up and down by **0.375** inches
To offset an entity, select the entity and then select
the *Offset* command. Enter the offset distance and
the options you desire.

Offset distance

Offsets in both directions

9) Change the middle line to a **Construction
Geometry**. To change an entity to a Construction
Geometry, right-click on it and then select the

Construction Geometry command

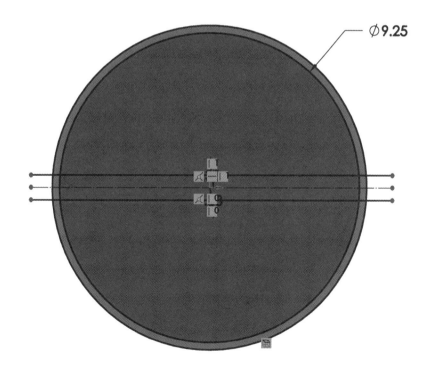

⌀9.25

> ➤ See section 3.1.3 to learn about ***Offsetting Entities***.
> ➤ See section 3.1.4 to learn about ***Construction Geometry***.

10) Use the **Trim to closest** command 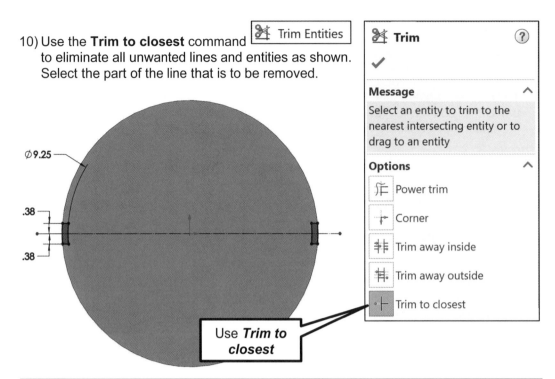 to eliminate all unwanted lines and entities as shown. Select the part of the line that is to be removed.

Use *Trim to closest*

> See section 3.1.1 to learn about *Trim* .command.

11) **Extrude** the sketch **0.75** inches.

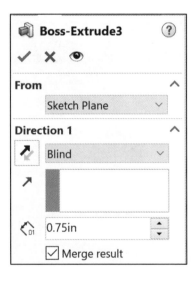

12) **Sketch** on the **Right Plane** and draw a **Circle** with the following sketch relations (see figure).

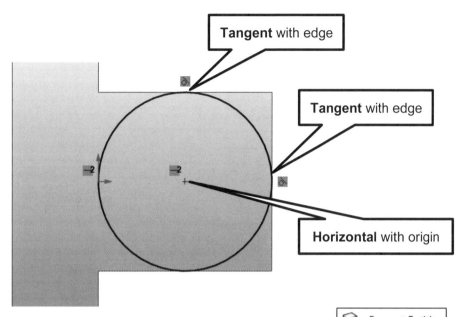

Tangent with edge

Tangent with edge

Horizontal with origin

13) Without leaving the sketch, use the **Convert Entities** command to make the top and right edges a sketch.

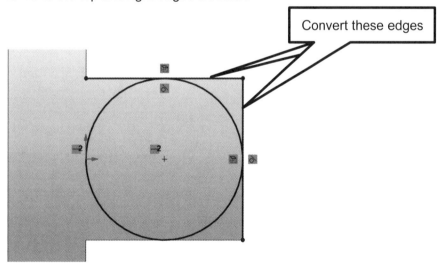

Convert these edges

14) Use the **Trim to closest** command 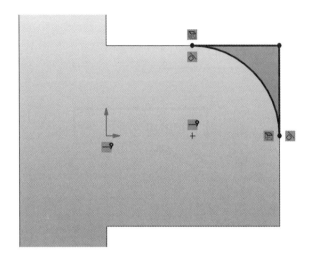 to eliminate all unwanted lines and entities as shown.

15) **Extrude Cut** the sketch **Through All – Both.**

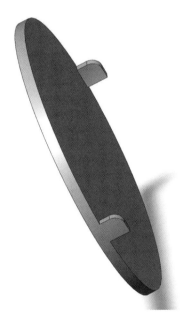

16) **Mirror** 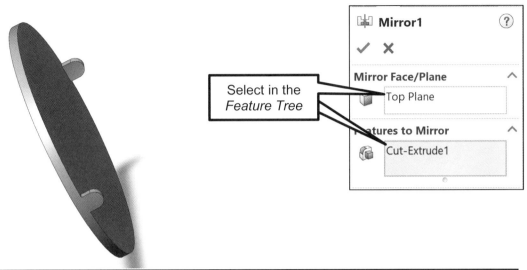, in the *Features* tab, the previous **Cut-Extrude** about the **Top Plane**. You need to expand your *Feature Design Manager Tree* and select the appropriate plane and feature.

> ➢ See section 3.2.4 to learn about **Mirror** .command.

17) Use the **Circle** 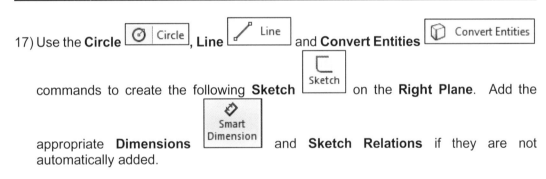, **Line** and **Convert Entities** commands to create the following **Sketch** on the **Right Plane**. Add the appropriate **Dimensions** and **Sketch Relations** if they are not automatically added.

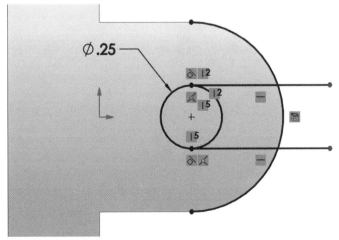

∅ .25

18) Use the **Trim to closest** command 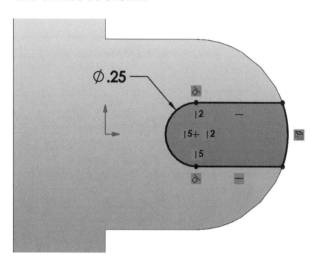 to eliminate all unwanted lines and entities as shown.

19) **Extrude Cut** the sketch **Through All – Both.**

20) **Fillet**  all edges shown as rounded in the figure to a radius of **0.1** inch.

21) Set your **Material** to **Chrome Stainless Steel.**

22) Evaluate the weight of your part [Mass Properties]. It should equal **4.07 lb**

23) **IMPORTANT!! Save** this part and keep it. It will be used in a later tutorial.

3.4.5) Setting up the Mirror Base

1) **Start SOLIDWORKS** and start a **new part** .

2) Set the drafting standard to **ANSI** and set the text to **upper case** for notes, tables, and dimensions. (*Options* ⚙ *– Document Properties – Drafting Standard*)

3) Set the units to **IPS** (inch, pound, second) and the **decimal = 0.123.** (*Options* ⚙ *– Document Properties – Units*)

4) Save the part as **MIRROR BASE.SLDPRT (File – Save)**. Remember to save often throughout this project.

3.4.6) Modeling the Mirror Base

1) **Sketch** on the **Front Plane**.

2) Draw, dimension, and apply the appropriate sketch relations to the sketch shown. Make sure that the sketch is constrained to the origin as shown.

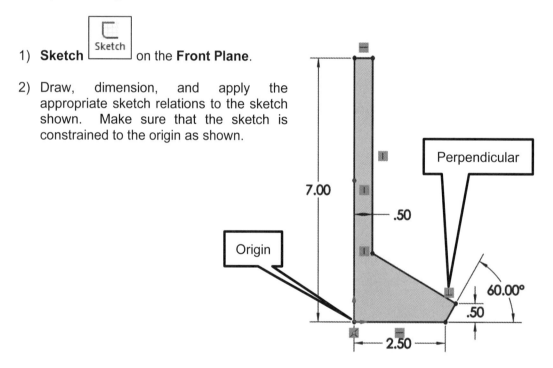

3) **Revolve Boss/Base**

 the sketch
360 degrees using the
centerline (**7.00" line**)
as the axis of
revolution and use the
Blind method.

4) Use the **Circle** [◎ Circle], **Line** [◢ Line] and **Trim to closest** [✄ Trim Entities]
commands to create the following **Sketch** [⌐ Sketch] on the **Front Plane**. Add the
appropriate **Dimension** [◇ Smart Dimension] and **Sketch Relations** if they are not automatically
added.

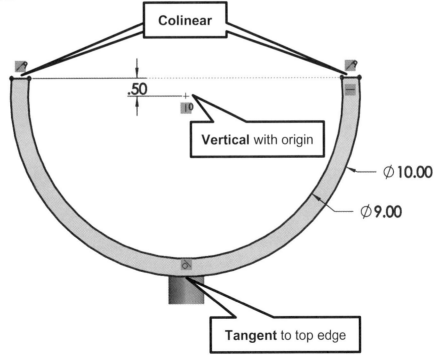

5) **Extrude**  about the **Mid Plane 1.00** inch.

Extruded
Boss/Base

⬢ **Boss-Extrude1**	?
✓ ✕ 👁	
From	⌃
Sketch Plane ⌄	
Direction 1	⌃
Mid Plane ⌄	
↗	
⟨ 1.00in	▲▼
☐ Merge result	

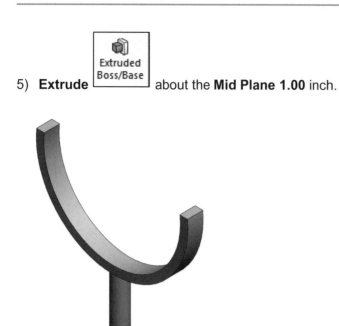

6) **Fillet** 🗋 Fillet the ends to a radius of **0.5** inch as shown in the figure.

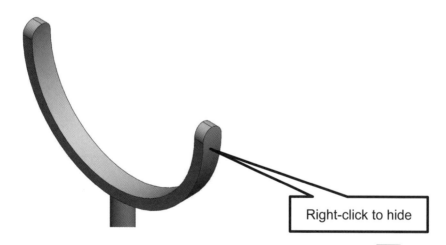

Right-click to hide

7) **Right-click** on the arms that hold the mirror and select **Hide** 🔖.

8) **Sketch** on the top end of the base and then **Convert Entities** .
Exit the **Sketch**.

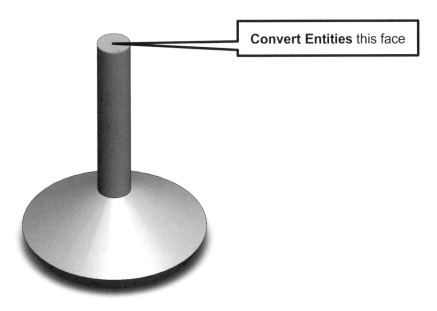

Convert Entities this face

9) Unhide the arms that hold the mirror by **right-clicking** the Boss-Extrude in the *Feature Design Tree* and selecting **Show** 👁.

10) **Extrude** 📦 Extruded Boss/Base the converted sketch **Up to Surface**.

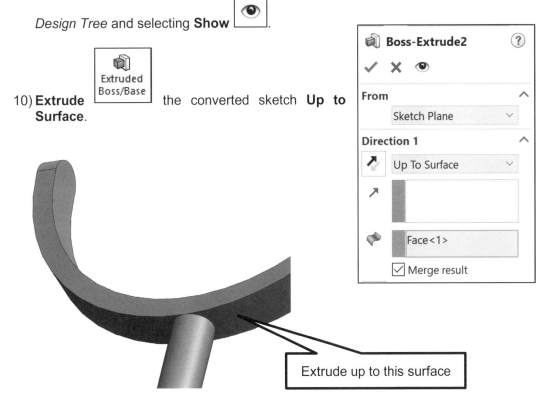

Extrude up to this surface

11) **Fillet** ⬡ Fillet where the base meets the arms using a **0.25**-inch radius as shown in the figure.

12) **Fillet** ⬡ Fillet the other edges of the base using a **0.1**-inch radius as shown in the figure.

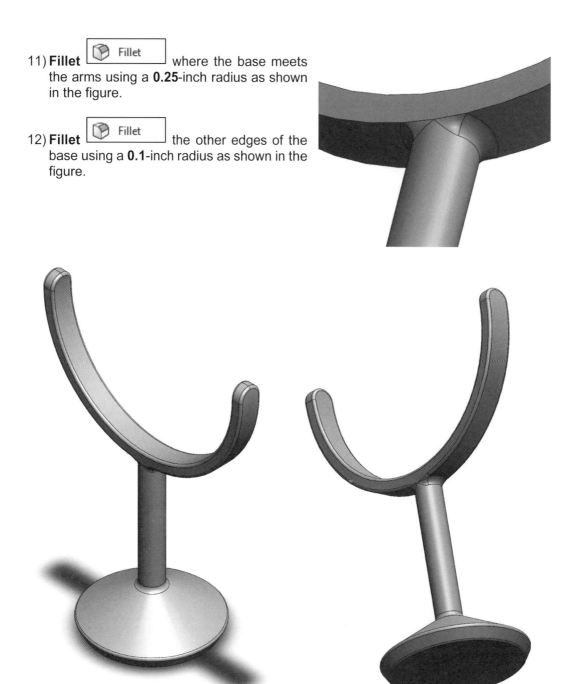

13) Use the **Circle** ⊘ Circle , **Line** ✏ Line , **Convert Entities** ⬚ Convert Entities and **Trim to closest** ✂ Trim Entities commands to create the following **Sketch** ⊏ Sketch on the **Front Plane**. Add the appropriate **Dimension** ◇ Smart Dimension and **Sketch Relations** if they are not automatically added.

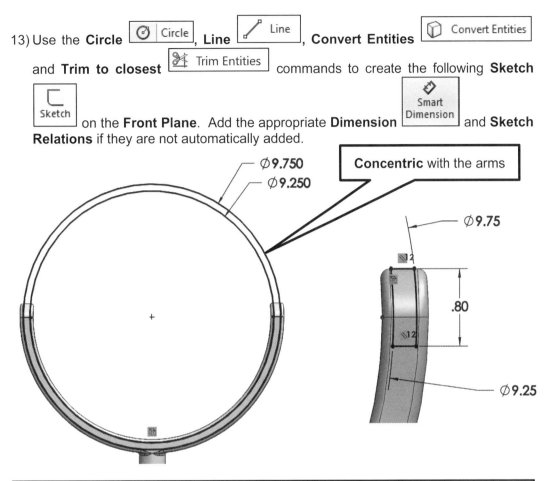

Ø**9.750**
Ø**9.250**

Concentric with the arms

Ø**9.75**

.80

Ø**9.25**

> Hint: You can use the *Convert Entities* to give a reference center point to draw the 9.75 and 9.25 diameter circles.

14) **Extrude Cut** 🗍 Extruded Cut the sketch **Through All – Both.**

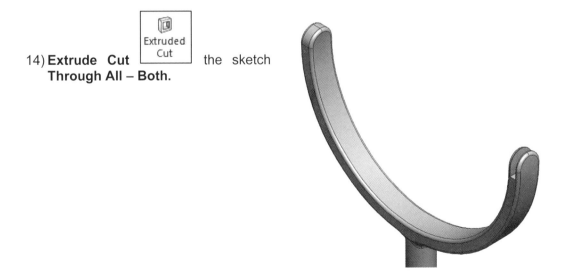

15) Create a reference **Plane** 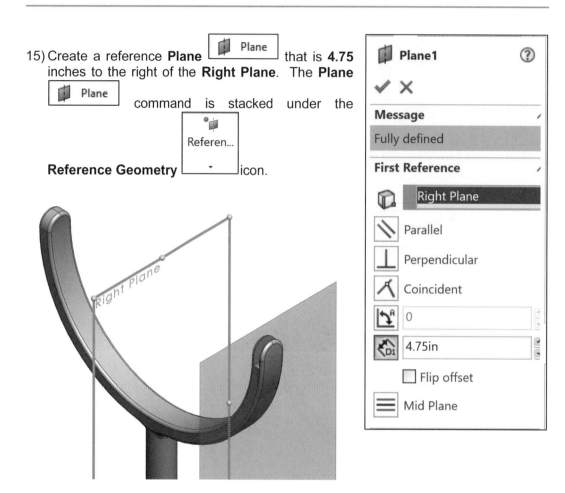 that is **4.75** inches to the right of the **Right Plane**. The **Plane** command is stacked under the **Reference Geometry** icon.

Plane1

Message
Fully defined

First Reference

Right Plane

Parallel

Perpendicular

Coincident

0

4.75in

Flip offset

Mid Plane

> See section 3.2.5 to learn about **Reference Geometries**.

16) Use the **Circle** command to create the following **Sketch** on the **new plane**. Add the appropriate **Dimension** **Smart Dimension**.

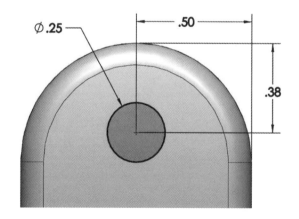

Ø.25   .50

.38

17) **Extrude**  **Up to Surface** in both directions.

18) **Mirror**  the previous **Cut-Extrude** and **Boss-Extrude** about the **Right Plane**.

19) **Hide Plane 1.**

20) Set your **Material** to **Chrome Stainless Steel.**

21) Evaluate the weight of your part [Mass Properties]. It should equal **10.00 lb**

22) **IMPORTANT!! Save** this part and keep it. It will be used in a later tutorial.

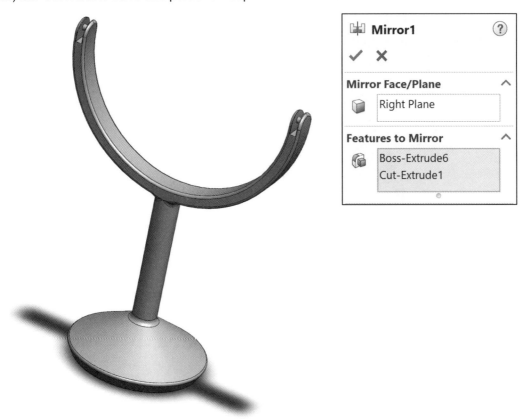

INTERMEDIATE PART MODELING IN SOLIDWORKS® PROBLEMS

P3-1) Create a solid model of the following 1020 Steel part and calculate the weight of the part. Dimensions are given in inches.

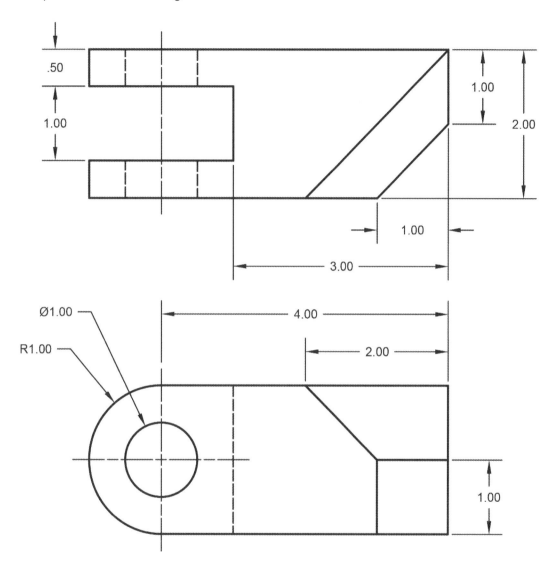

P3-2) Create a solid model of the following 1060 Alloy Aluminum part and calculate the weight of the part. Dimensions are given in millimeters.

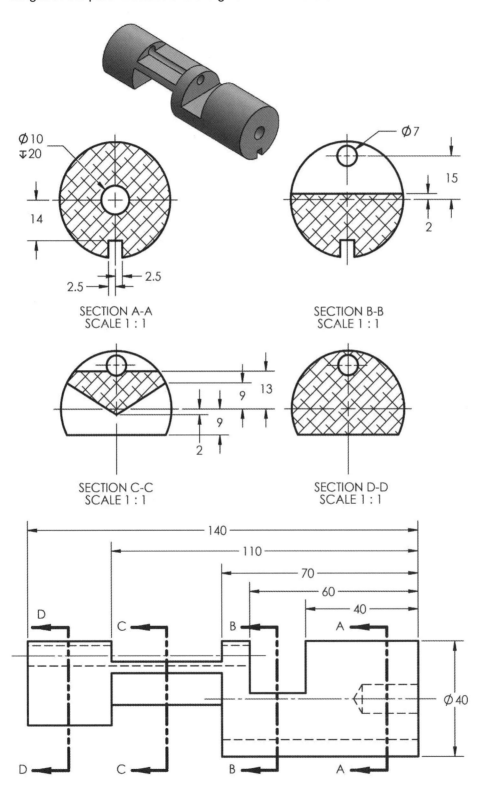

Ø10
⫂20

14

2.5 — 2.5

SECTION A-A
SCALE 1 : 1

Ø7

15

2

SECTION B-B
SCALE 1 : 1

13
9

9

2

SECTION C-C
SCALE 1 : 1

SECTION D-D
SCALE 1 : 1

140

110

70

60

40

D

C

B

A

Ø40

D

C

B

A

P3-3) Create a solid model of the following 1060 Alloy Aluminum **JIG**. Dimensions are in millimeters.

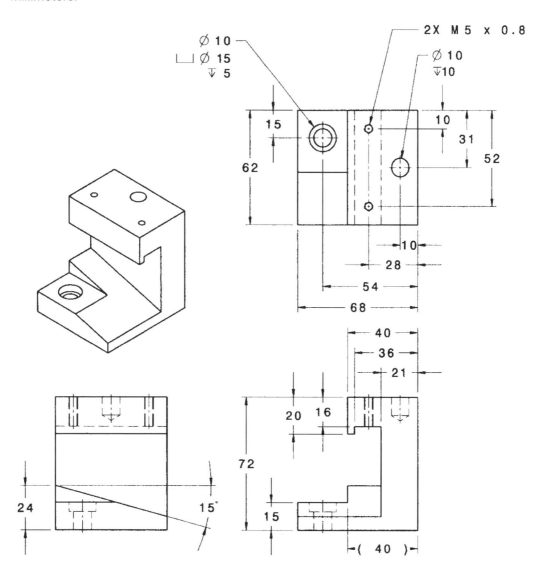

P3-4) Create a solid model of the following Tool Steel **FIXTURE**. Dimensions are in millimeters.

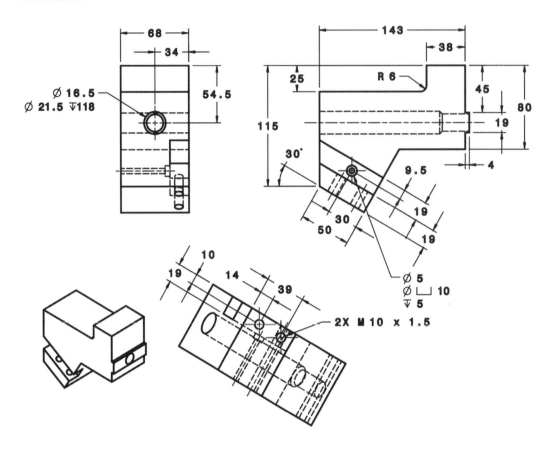

P3-5) Create a solid model of the following Grey Cast Iron **PULLEY**. Dimensions are in millimeters. All fillets and rounds R2 unless otherwise specified.

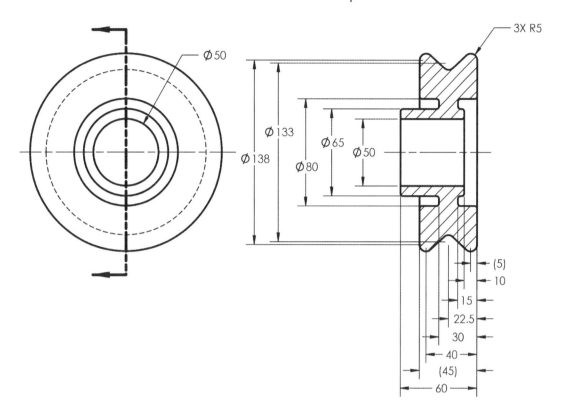

P3-6) Create a solid model of the following High Density Polyethylene **STEP GEAR MECHANISM**. Dimensions are in millimeters.

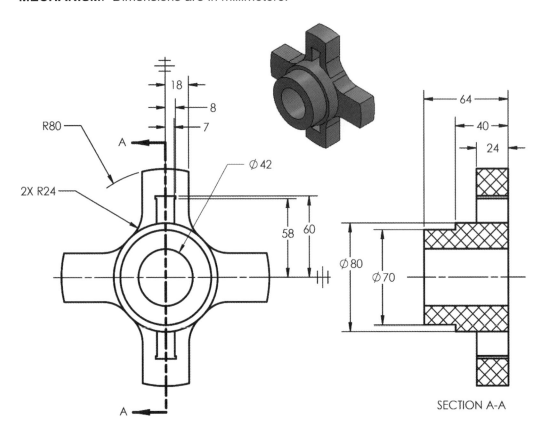

SECTION A-A

P3-7) Use SOLIDWORKS® to create a solid model of the following PVC Rigid part and record the mass. Note that the draft angle 'B' changes and may be adjusted in the Extrude command. Dimensions are given in millimeters.

	Group 1	Group 2	Group 3
B	10	8	12

Mass = _____ grams

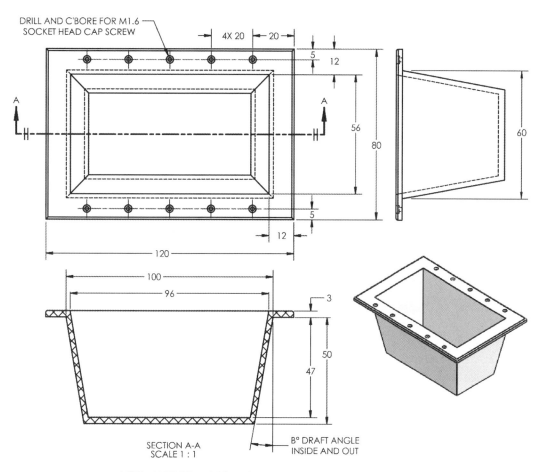

DRILL AND C'BORE FOR M1.6 SOCKET HEAD CAP SCREW

4X 20   20

5   12

56

80

60

5

12

120

100

96

3

50

47

SECTION A-A
SCALE 1 : 1

B° DRAFT ANGLE
INSIDE AND OUT

NOTE: ALL FILLETS AND ROUNDS
R1 UNLESS OTHERWISE SPECIFIED

NOTES:

CHAPTER 4

INTERMEDIATE DRAWINGS IN SOLIDWORKS®

CHAPTER OUTLINE

4.1) CREATING DRAWINGS IN SOLIDWORKS®..2

4.2) DRAWING VIEWS...2

    4.2.1) Section views 2

    4.2.2) Detail views 4

4.3) ANNOTATIONS...5

    4.3.1) Leader Notes 5

    4.3.2) Datum Feature 6

4.4) SKETCHING ON A DRAWING 7

4.5) CONNECTING ROD PRINT TUTORIAL..8

    4.5.1) Prerequisites 8

    4.5.2) What you will learn...8

    4.5.3) Setting up the drawing views 10

    4.5.4) Creating a detail view 15

    4.5.5) Adding dimensions...17

    4.5.6) Filling in the title block..22

4.6) FLANGED COUPLING PRINT TUTORIAL 24

    4.6.1) Prerequisites 24

    4.6.2) What you will learn?...24

    4.6.3) Creating the views 25

    4.6.4) Adding dimensions...28

CREATING INTERMEDIATE DRAWINGS IN SOLIDWORKS® PROBLEMS 33

CHAPTER SUMMARY

In this chapter, you will learn how to create special types of projected views in SOLIDWORKS®. Specifically, you will learn how to create section views and detailed views. However, problems at the end of the chapter will take you through creating other types of views. By the end of this chapter, you will be able to create several different types of projected views.

4.1) CREATING DRAWINGS IN SOLIDWORKS®

It is very simple to create an orthographic projection of a part in SOLIDWORKS®. Once a solid model is created, it just takes a few clicks to create an orthographic projection and pictorial of the model. However, sometimes you need to add special projected views in order to convey a complete picture of your part. So, it is your job to inspect the part print, add any needed views, and make any adjustments to the drawing that are required so that all the necessary information is conveyed.

4.2) DRAWING VIEWS

When a new drawing is started, you select the model file that you wish to create the orthographic projection from and then you select the views you want to show. After the drawing or paper space has been entered, additional views can be added using the commands located in the *View Layout* tab (see Figure 2.5-1). Projected views, auxiliary views, section views, and detail views are some of the options that are available to be added here.

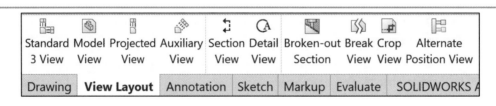

Figure 4.2-1: View Layout tab

4.2.1) Section views

The **Section View** command is located in the *View Layout* tab and also in the *Drawing* tab. A section view allows a view from inside of the part. An imaginary cut is made through the part and a portion of the part is mentally removed allowing for this view. The imaginary cut is indicated on a part print using a **Cutting Plane line**. The material that is actually cut is indicated with **Section lines**. The section line pattern is dependent on the material of the part. This is automatically chosen by the program. To create a section view, use the following steps.

1) Select the **Section View** command in the *View Layout* tab of the *Command Manager* ribbon.
2) Select the type of section that is to be made and name the section view (see Figure 4.2-1). Note, the first section view is normally named A.
3) Place the cutting plane line by snapping to the appropriate point or points. Note that more points may be added in order to bend the cutting plane line. Getting an offset section cutting plane line just right may take a little practice.
4) Select OK when the section line is what is wanted.
5) Place the section view.
6) Click on the section view and remove all hidden lines by selecting **Hidden Lines Removed** as the *Display Style* (see Figure 4.2-1).
7) The alignment of the view can be broken by right-clicking on the section view and selecting **Alignment – Break Alignment** to create a removed section.

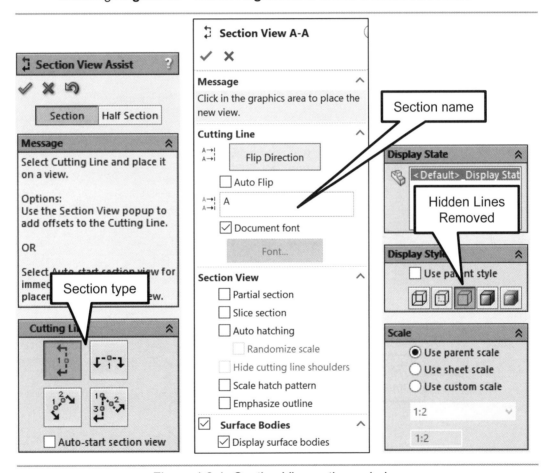

Figure 4.2-1: *Section View* options window

4.2.2) Detail views

The **Detail view** ![Detail View icon] command is in the *View Layout* tab and also in the *Drawing* tab. A detail view is a partial view of a portion of the part shown at an increased scale. Use the following steps to create a detail view.

1) Select the **Detail View** ![Detail View icon] command located in the *View Layout* tab.
2) Draw a circle around the feature where a detail view is wanted.
3) Place the detail view.
4) Properties of the detail view may be changed in the *Detail View* options window (see Figure 4.2-2).
5) Adjust the detail circle so that the letter can be read from the bottom of the drawing. This can be done by clicking and dragging the letter.
6) Click on the detail view and adjust the scale as needed.

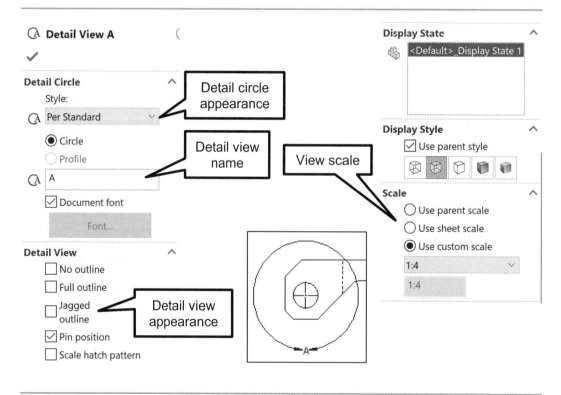

Figure 4.2-2: *Detail View* options window

4.3) ANNOTATIONS

Drawings may be annotated by using the commands located in the **Annotation tab**. Annotation commands include adding dimensions, notes, surface finish symbols, GDT notes and symbols, centerlines, and center marks. The *Annotation tab* is shown in Figure 4.3-1.

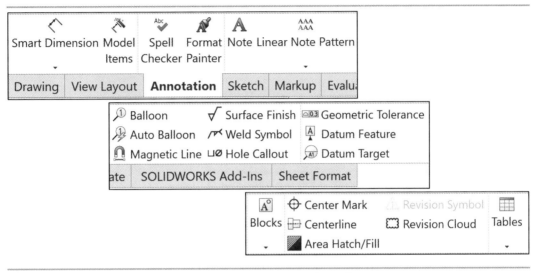

Figure 4.3-1: Annotation tab

4.3.1) Leader Notes

Leader lines with notes can be added to your drawing using the **notes** command which is located in the *Annotation* tab. Different options and text formatting are available. Figure 4.3-2 shows the *Note* option window and the text *Formatting* window.

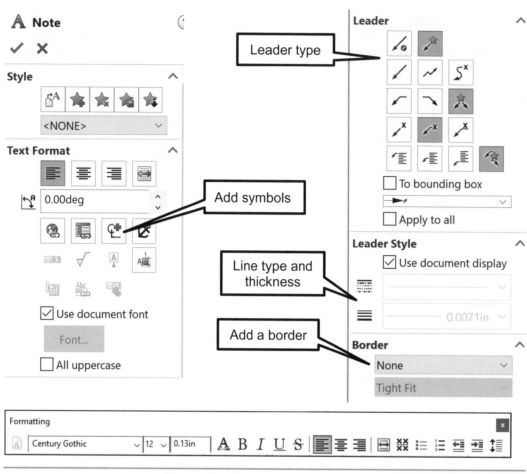

Figure 4.3-2: *Note* option window

4.3.2) Datum Feature

A **datum feature** is a functionally important surface on a part. There are usually three datum features per part. Most of the time, features on a part are located with respect to these features. Usually, datum features are the origin of all the dimensions. Datum features on a drawing are identified by a datum feature symbol. Datum feature symbols are created using the **Datum Feature** [A] Datum Feature command located in the *Annotation* tab. A letter is used to identify and differentiate between the datum features. Figure 4.3-3 shows the use of datum features on a drawing.

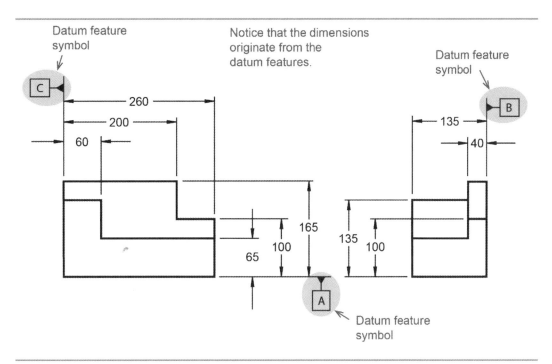

Figure 4.3-3: Dimensioning using datum features.

4.4) SKETCHING ON A DRAWING

A drawing may be sketched on if additional features need to be added. The *Sketch* tab in a drawing contains many of the same commands as the *Sketch* tab in a part file. The *Sketch* tab is shown in Figure 4.4-1.

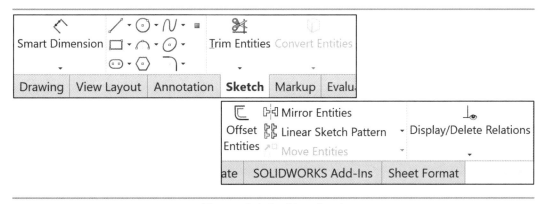

Figure 4.4-1: Sketch tab

4.5) CONNECTING ROD PRINT TUTORIAL

4.5.1) Prerequisites

To complete this tutorial, the user should have completed the listed tutorial and be familiar with the listed topics.

- Chapter 1 – Connecting Rod tutorial
- Passing familiarity with orthographic projection.
- Ability to read dimensions.

4.5.2) What you will learn

The objective of this tutorial is to introduce you to the SOLIDWORKS' drawing capabilities (i.e., the ability to create orthographic projections.) In this tutorial, you will be creating a part print of the connecting rod that you modeled in Chapter 1. The part print is shown in Figure 4.5-1. Specifically, you will be learning the following commands and concepts.

Drawing

- Detail view
- Leader Note
- Moving and Editing Dimension
- Datum Feature

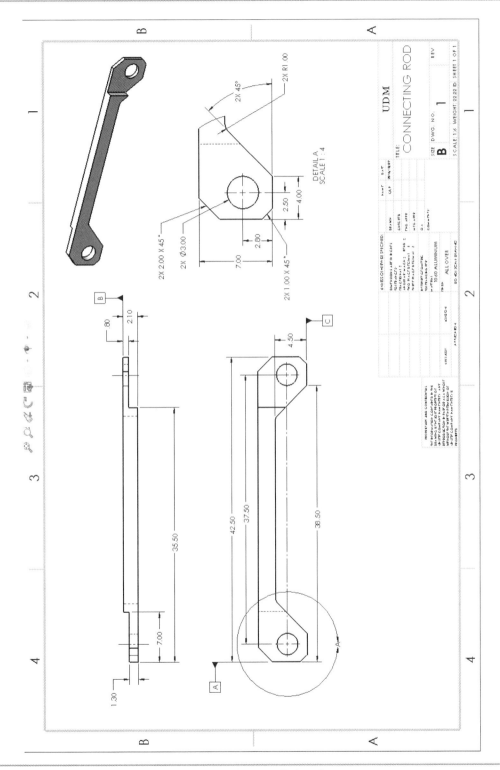

Figure 4.5-1: Detail drawing

4.5.3) Setting up the drawing views

It is often the case that CAD programs do not follow the ASME drawing standards when creating a part print. Therefore, it is important that user reviews the print and decides what is correct and changes what needs to be adjusted.

1) **Open CONNECTING ROD.SLDPRT**.

2) Start a **New Drawing** Drawing.

3) Select a **Sheet Size** of **B**. If only the metric sheet sizes are shown, deselect the *Only show standard formats* check box.

4) Set the drafting standard to **ANSI** and set the text to **upper case** for notes, tables, and dimensions. (*Options* ⚙ – *Document Properties* – *Drafting Standard*)

5) Set the units to **IPS** (i.e. inch, pound, second) and set the **Decimals = .12**. Also, select the rounding option, **Round half to even**. (*Options* ⚙ – *Document Properties* – *Units*)

6) **Save** the drawing as **CONNECTING ROD.SLDDRW**   Remember to **Save** often throughout this tutorial.

Model
View

7) In the **View Layout** tab, select **Model View**.
   In the *Model View* window, double click on your
   CONNECTING ROD part and then
   a. Select **Create multiple views**.
   b. Choose to create a **Front** and **Top** view. Deselect
      the Isometric view if it is selected.
   c. Select **Trimetric** for the pictorial type.
   d. Select a display style = **Hidden Lines Visible**.
   e. Select the **Use custom scale** radio button and
      enter a scale of **1:6**.

   f. ✓

➢ Note: Views can always be added later using the
  Projected View command.
---

8) Select the pictorial view in the drawing area and set the
   display style as **Shaded with Edges** and use a **1:8**
   scale.

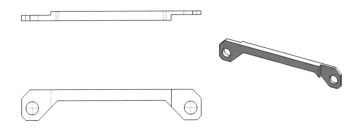

9) Move the views so that they are completely inside the
   drawing border. This is done by clicking and dragging
   on the view.

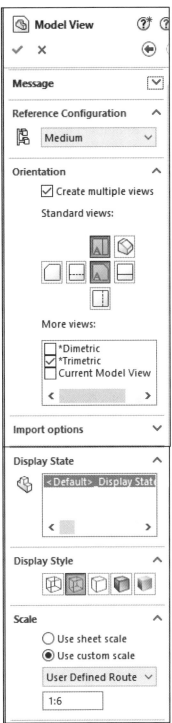

10) Add missing **Centerlines**. Click on **Centerlines** 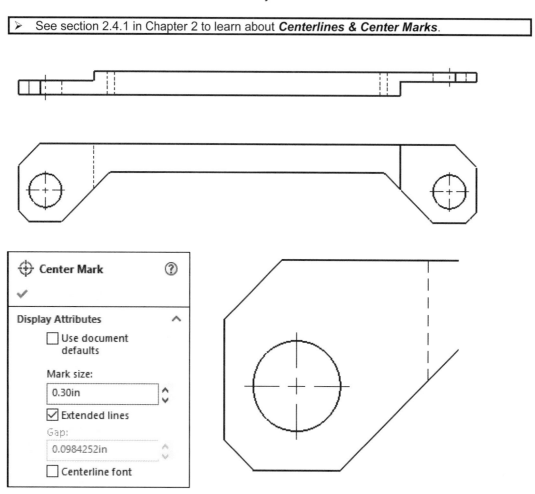 in the *Annotation* tab. Select the **Select View** check box, and then select the **top view**. Click **OK** in the *Centerline* window.

11) Adjust the **Center Marks** properties to increase the size of the short dash to **0.3 in**. In the front view click on the center mark. Deselect the **Use document defaults** check box and adjust the *Mark size* to **0.3in**.

> ➢ See section 2.4.1 in Chapter 2 to learn about ***Centerlines & Center Marks***.

12) If the *Layers* toolbar is not showing (it should be in the bottom left corner), activate it (**View – Toolbars – Layer**).

13) Create a **New** layer called **Centerline**. Do this by clicking **Layers Properties**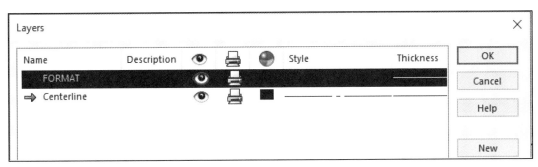
and then selecting **New**. Name the layer and set the line *Style* to **Center**. Select **OK**.

> ➤   See section 2.5 in Chapter 2 to learn more about *Layers*.

14) Set the current layer (i.e. the layer you will be drawing on) to
Centerline and then *Sketch* a **Line** between the
two center marks indicating that the holes are in line. To get
an accurate line, snap to the quadrants of the circles.

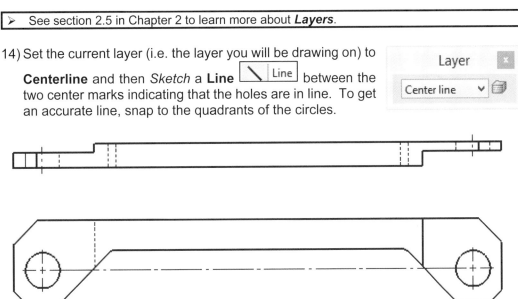

15) Notice the two hidden lines in the top view that indicate a change in surface direction
of the fillets. We want to hide these hidden lines. Right click on one of the inner hidden

lines and select **Hide/Show Edges** . Repeat for the other inner hidden line.

16) Phantom lines are used to show changes in surface directions. Right click on the remaining hidden lines and change the line type to **Phantom**. It may not look much different than the hidden line type. This is because it only traverses a short distance. If the lines were longer, they would look more like phantom lines.

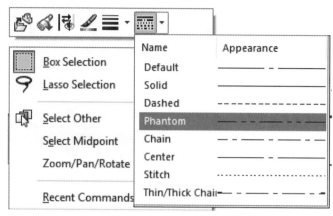

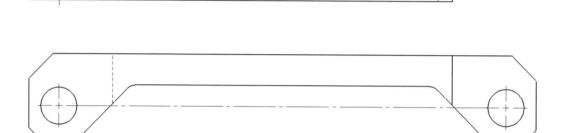

17) **Save.** A *Save* window may appear wanting you to save all models associated with this drawing. Select **Save All**.

4.5.4) Creating a detail view

1) Drag the views to the approximate locations shown.

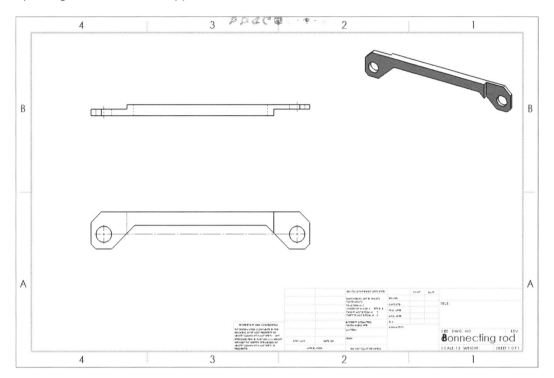

2) Change the current layer to **-None-** and then create a **1:4** scale **detail view** of the left end of the connecting rod. Adjust the *Center Mark* properties to increase the center dash length.

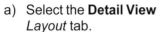

a) Select the **Detail View** 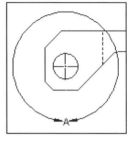 command located in the *View Layout* tab.
b) Draw a circle around the left end of the connecting rod using the center of the hole as the center.
c) Place the detail view to the right of the front view.
d) Adjust the detail circle so that the letter can be read from the bottom of the drawing. This can be done by clicking and dragging the letter.
e) Click on the detail view and adjust the scale to **1:4**.
f) Click on the center mark in the detailed view and increase the length of the dash.

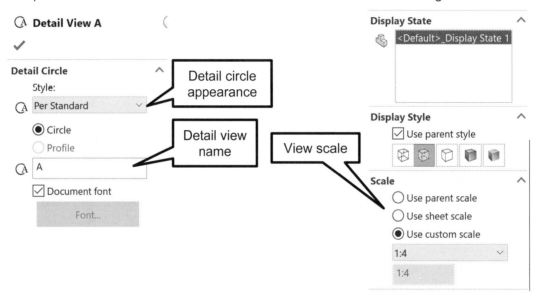

> ➢ For information on *Detail Views* see section 4.2.2.

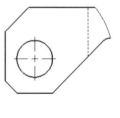

DETAIL A
SCALE 1 : 4

4.5.5) Adding dimensions

1) Add **Model Items** (*Annotation* tab) to the drawing using the following settings. Don't worry about the fact that the dimensions are all jumbled and overlapping. We will move them into more appropriate locations.
   a) Source = **Entire model**
   b) Activate the **Import items into all views** check box.
   c) Dimensions = **Marked for drawing**
   d) Activate the **Eliminate duplicates** check box.
   e) Set the Layer = **None**

> ➢ For information on **Model Items** see Chapter 2

2) We want to make the top view look like the figure shown below.

2a) This figure below shows how my top view ended up. Note that if any of your dimension text is vertical, you are in the ISO standard. Change that to the ANSI standard (***Options*** icon – ***Document Properties*** tab – ***Drafting Standard - ANSI***). If you are missing any dimension, try to locate it in another view. If you find it, hold the **SHIFT** key down and drag the dimension to the top view.

2b) Drag the dimensions and the dimension text into the correct locations. Holding the **ALT** key will produce a smooth drag.

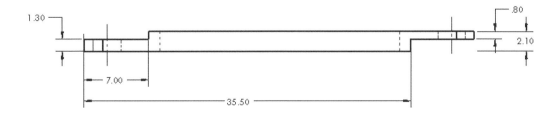

2c) Check and see if all the dimensions have a gap between the part and the extension line. Not all the dimensions will. Click on the dimension and then drag the appropriate grip box to produce a gap. You may also find dimensions that have too big of a gap. Check the figure at the beginning of this step to see how the dimensions should eventually look.

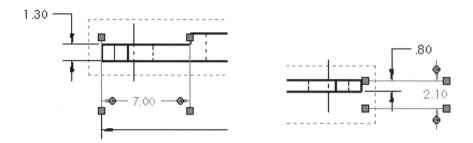

3) Make the Front view look like the following figure. Remember, if the dimension is in another view, hold the **SHIFT** key and drag the dimension to the front view. Remember to adjust the gaps of the extension lines. Note that the chamfer dimensions are given using a linear dimension and an angle. Delete the dimensions associated with the following chamfer (1.00 x 45°, 2.00 x 45°). We will be adding those back using a leader line and note. If the R1.00 dimension is in the Front view, try to drag it to the Detail view. If it doesn't work, just delete the dimension. We will put it back later.

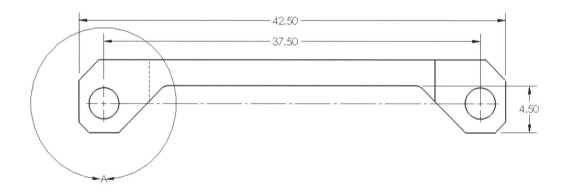

4) After making the appropriate adjustments, the detail view should look like the figure shown. Your view will look very messy now. Start by dragging the dimension away from the part.

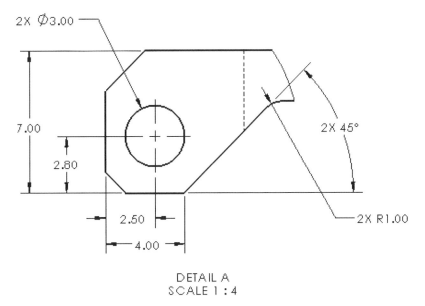

DETAIL A
SCALE 1 : 4

4a) **Delete** the dimensions associated with the (1.00 X 45°) and (2.00 X 45°) chamfer if they are included in the detail view. The following figure shows how the view should end up after adjusting the dimension positions. The radius dimension may look different or not be there at all. Don't worry about that now. Adjust the gaps as needed.

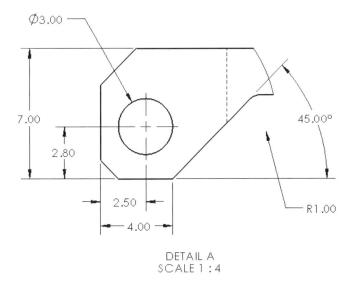

DETAIL A
SCALE 1 : 4

4b) Add the **R1.00** dimension in the detail view if yours is missing. Use the **Smart Dimension** command located in the *Sketch* tab. Notice that the dimension is gray. We will take care of that later.

4c) Click on the R1.00 dimension in the detail view and add **2X** to the dimension text and then set the following leader properties by clicking on the **Leader** tab **(Witness/Leader Display = Outside, Radius, Open Leader)**.

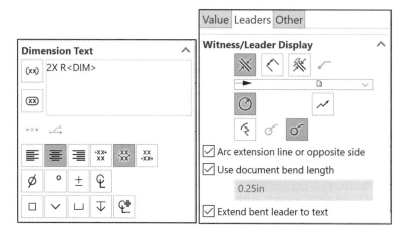

4d) Click on the 45.00° dimension in the detail view and add **2X** and change the *Precision* to **None** if there are 2 decimal places.

4e) Click on the ⌀3.00 dimension in the detail view and add **2X** to the dimension text.

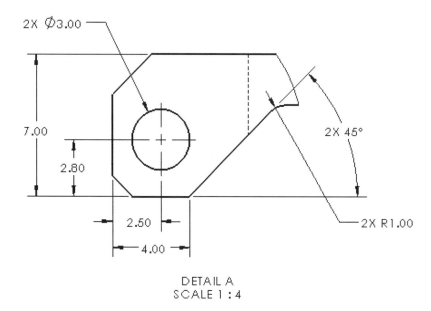

DETAIL A
SCALE 1 : 4

5) Create a **Dimension** layer with a **Solid** line type and set it to be the current layer.

6) Add the following 38.50 dimensions to the front view using **Smart Dimension**. Make sure that it is on the **Dimension** layer. If the dimensions look gray, it is not on the correct layer.

7) Click on the R1.00 dimension in the Detail view and place it on the *Dimension* layer if it is gray.

8) Add the chamfer dimensions to the detail view in the *Dimension* layer. Select the **Note**

A
Note command located in the *Annotation* tab and then select the chamfer edge. Pull the leader out and away from the view, click and then add the text. To add the degree symbol, click on **Add Symbol** .

> ➤ To learn more about *Notes*, see section 4.3.1.

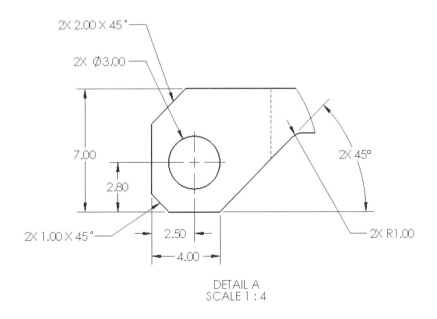

DETAIL A
SCALE 1 : 4

9) Add **Datum Feature** symbols 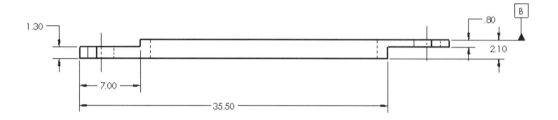 (*Annotation* tab) as shown in the figure. Do this by selecting the extension line that you wish to attach the datum feature symbol to and then pull it out to the desired position.

> ➢   To learn more about *Datum Features*, see section 4.3.2.

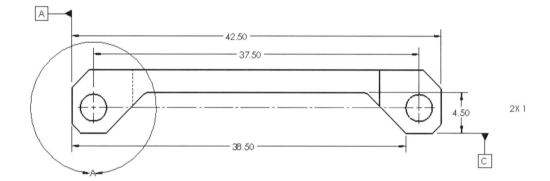

4.5.6) Filling in the title block

1) **Right-click** on the drawing somewhere outside the views but inside the drawing area and select **Edit Sheet Format**. Notice that the drawing will disappear.

2) Move the mouse around the title block and notice that every once in a while a text symbol will appear. This indicates a text field. To enter text, just double click on the field.

TITLE:

A

3) Enter the following text. Note that all text should be capitalized. The formatting window can be used to increase or decrease the text size as needed.
   a) TITLE = **CONNECTING ROD**
   b) Above the title place the school's name.
   c) DWG. NO. = **1**
   d) REV = **1**
   e) WEIGHT = Enter the value given from the *Mass Properties*
   f) DRAWN NAME = your initials
   g) DRAWN DATE = enter the date in this format (YYYY/MM/DD)
   h) MATERIAL = **1060 ALUMINUM**
   i) FINISH = **ALL OVER**

4) Get out of your title block by **right-clicking** on your drawing and selecting **Edit Sheet**.

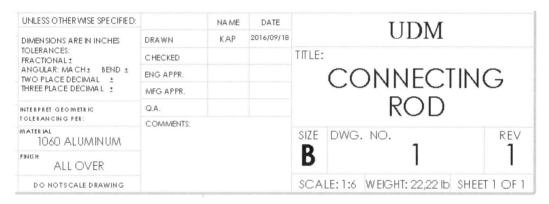

4.6) FLANGED COUPLING PRINT TUTORIAL

4.6.1) Prerequisites

Before starting this tutorial, the user should have completed the following tutorials.

- Chapter 2 - Angled Block print tutorial
- Connecting Rod print tutorial

It will help if the user has the following knowledge.

- A familiarity with section views.
- A familiarity with threads and fasteners.

4.6.2) What you will learn?

The objective of this tutorial is to continue working with SOLIDWORKS' drawing capabilities. In this tutorial, you will be creating a part print of the *Flanged Coupling* that was modeled in Chapter 3. The part print is shown in Figure 4.6-1. Specifically, you will be learning the following commands and concepts.

Drawing

- Section view
- Sketching on a drawing

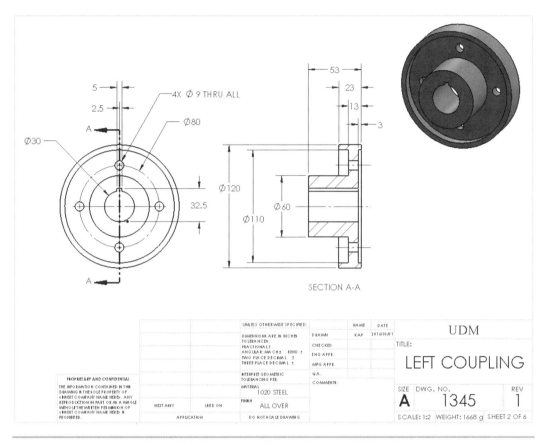

Figure 4.6-1: Flanged Coupling detail drawing

4.6.3) Creating the views

1) **Open COUPLING.SLDPRT** that was created in Chapter 3.

2) Start a **New Drawing** Drawing of **Sheet Size** of **A (ANSI) Landscape**.

3) Set up the following views of the part. Use the sheet scale of **1:2**.

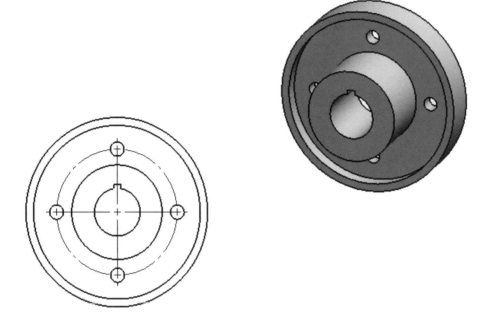

4) Set the drafting standard to **ANSI** and set the text to **upper case** for notes, tables, and dimensions.

5) Set the units to **MMGS** (millimeters, grams, second) and the **decimal = 0.1.**

6) Create a **Section View** of the part.

a) Select **Section View** in the *View Layout* tab.
b) Cutting Line = **Vertical**
c) Select the center of the part and then select the green check mark.
d) Pull the section view to the right of the Front view and click the left mouse button.

➢ To learn more about *Section Views*, see section 4.2.1.

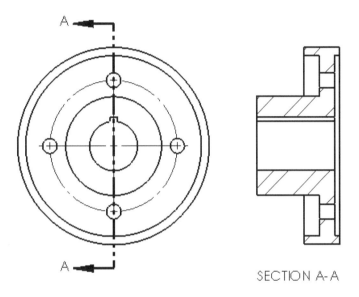

SECTION A-A

7) Add missing **Center lines** [Centerline] to the section view. Extend the centerlines as necessary. We do not want the center lines to end at the boundary of the part.

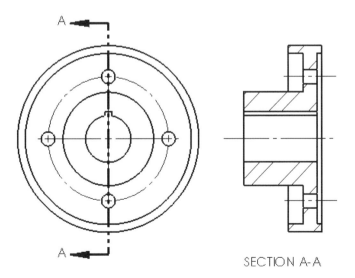

SECTION A-A

4.6.4) Adding dimensions

1) Add **Model Items** [Model Items icon] to the drawing using the following settings.
   - Source = **Entire model**
   - Activate the **Import items into all views** check box.
   - Dimensions = **Marked for drawing, Hole Wizard locations,** and **Hole callouts**
   - Activate the **Eliminate duplicates** check box.
   - Set the Layer = **None**

2) Move the dimensions away from the part views so that they are not directly on top of the view. Note that many of these dimensions are not ideal. We will need to clean up the dimensions. Don't worry if your dimensions are not exactly what is shown, the *Model Items* are just a place to start.

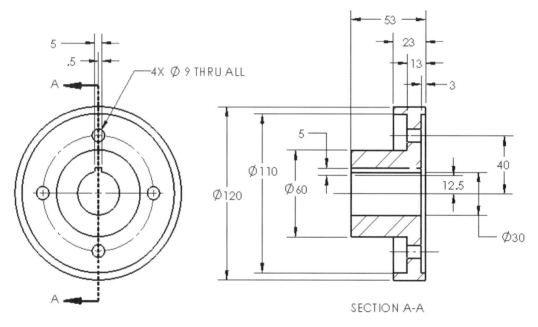

SECTION A-A

3) The 4X ∅9 or ∅9 THRU dimension is gray. Create a **Dimension** layer and move that dimension to this layer. Do this by clicking on the dimension and then selecting the *Dimension* layer.

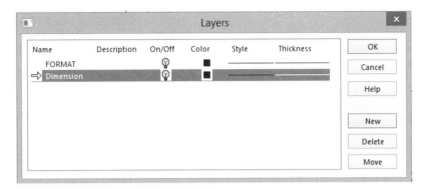

4) **Delete** the 12.5 mm and 5 mm dimension in the section view that dimension the keyway and dimension the keyway in the front view. Place all new dimensions on the **Dimension** layer. To add this dimension, you first need to draw a short horizontal line (shown in the figure). Make this dimension a driven dimension if asked.

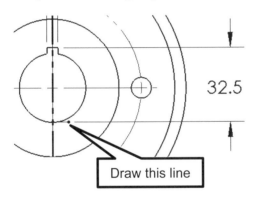

Draw this line

5) Delete the ∅30 and 40 radius dimensions from the section view. We will be adding these dimensions to the front view.

6) Replace the diameter dimensions in the front view (i.e., ⌀30 and ⌀80) using the **Smart Dimension** command.

- If the ⌀30 dimension wants to be a radius, click on the *Leader* tab and select **Diameter**.
- If you are having trouble dimensioning the circle of centers ⌀80, you may need to delete this circle and draw one in manually. If this is the case, make a new layer called **Centerline** using a centerline line type and place the new circle on this layer. Make the dimensions driven if asked.

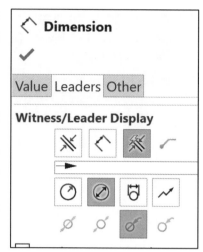

7) Change the ⌀9 dimension text to what is shown.

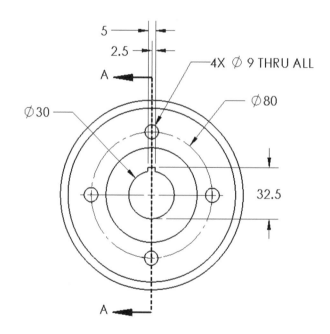

8) The drawing below is how the final views should look. If they do not, make any adjustments necessary.

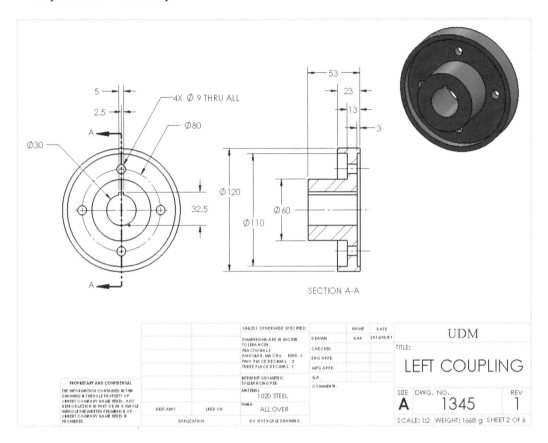

9) Fill in the title block with the following information.
   a) TITLE = **LEFT COUPLING**
   b) Above the title place your school's name.
   c) DWG. NO. = **1345**
   d) REV = **1**
   e) WEIGHT = Determine the weight of your part.
   f) SHEET = 2 OF 6
   g) DRAWN NAME = your initials
   h) DRAWN DATE = enter the date in this format (YYYY/MM/DD)
   i) MATERIAL = **1020 STEEL**
   j) FINISH = **ALL OVER**

NOTES:

CREATING INTERMEDIATE DRAWINGS IN SOLIDWORKS® PROBLEMS

P4-1) Use SOLIDWORKS® to create the part print that looks exactly like the detailed drawing of the model shown in **P1-13**. Fill in the appropriate information into your title block. Enter the weight of your part into the title block of your drawing.

P4-2) Use SOLIDWORKS® to create the part print that looks exactly like the detailed drawing of the model shown in **P1-14**. Fill in the appropriate information into your title block. Enter the weight of your part into the title block of your drawing.

P4-3) Use SOLIDWORKS® to create the part print that looks exactly like the detailed drawing of the model shown in **P3-2**. Fill in the appropriate information into your title block. Enter the weight of your part into the title block of your drawing.

P4-4) Use SOLIDWORKS® to create the part print that looks exactly like the detailed drawing of the model shown in **P3-3**. Fill in the appropriate information into your title block. Enter the weight of your part into the title block of your drawing.

P4-5) Use SOLIDWORKS® to create the part print that looks exactly like the detailed drawing of the model shown in **P3-4**. Fill in the appropriate information into your title block. Enter the weight of your part into the title block of your drawing.

P4-6) Use SOLIDWORKS® to create the part print that looks exactly like the detailed drawing of the model shown in **P3-5**. Fill in the appropriate information into your title block. Enter the weight of your part into the title block of your drawing.

P4-7) Use SOLIDWORKS® to create the part print that looks exactly like the detailed drawing of the model shown in **P3-6**. Fill in the appropriate information into your title block. Enter the weight of your part into the title block of your drawing.

NOTES:

CHAPTER 5

CONFIGURATIONS IN SOLIDWORKS®

CHAPTER OUTLINE

5.1) WHAT IS A CONFIGURATION? 2

    5.1.1) Adding a Configuration 2

    5.1.2) Design Table.. 3

5.2) CONNECTING ROD CONFIGURATION TUTORIAL.. 4

    5.2.1) Prerequisites 4

    5.2.2) What you will learn.. 4

    5.2.3) Creating configurations 5

5.3) FLANGED COUPLING CONFIGURATION TUTORIAL...................................... 8

    5.3.1) Prerequisites 8

    5.3.2) What you will learn.. 8

    5.3.3) Creating Configurations 9

5.4) RIGID COUPLING DESIGN TABLE TUTORIAL.. 10

    5.4.1) Prerequisites 10

    5.4.2) What you will learn.. 10

    5.4.3) Setting up the project.. 11

    5.4.4) Creating the base configuration 11

    5.4.5) Setting up the design table 15

    5.4.6) Suppressing features and changing material 22

CONFIGURATIONS IN SOLIDWORKS® PROBLEMS... 25

CHAPTER SUMMARY

In this chapter, you will learn how to create configurations in SOLIDWORKS®. A configuration of a part is essentially the part with something changed. For example, the size (e.g., length of a bolt), a feature (e.g., added holes), or the material. It is used when not enough needs to be changed to model a brand-new part. By the end of this chapter, you will be able to create configurations of an existing part.

5.1) WHAT IS A CONFIGURATION?

A configuration is a variation of a part. Basically, it is the same part, but in a different size or with a minor change in a feature or material. Not enough is different about the part that would warrant you modeling a completely different part. Configurations make it easy to go back and forth between versions of the part. For example, Figure 5.1-1 shows a bolt in two different lengths or configurations.

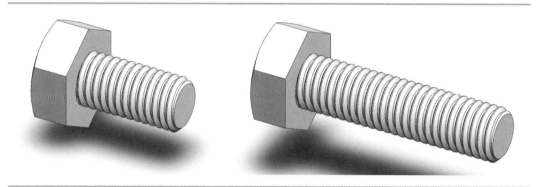

Figure 5.1-1: Configurations of a bolt

5.1.1) Adding a Configuration

A configuration is added after a base part model has been created. Use the following steps to add a configuration. Figure 5.1-2 shows the windows associated with these steps.

Adding a Configuration

1) Click on the **Configurations Manager** tab ⬚.
2) **Right-click** on the root part and select **Add Configuration…**
3) Name the configuration.
4) Add a description.
5) Click ⬚.
6) If a warning window appears, select **Yes**.

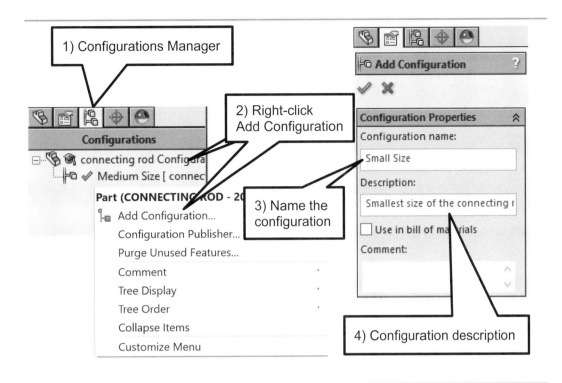

Figure 5.1-2: Adding a Configuration

5.1.2) Design Table

Design tables allow for multiple configurations of a part (or assembly) to be built all at once. These parameters are built in a Microsoft® Excel spreadsheet. To use design tables, Microsoft® Excel must be installed on the computer. Listed are some of the things that can be controlled in a part design table. There are more than what are listed, but these are the most common.

- Dimensions
- Suppression states of features
- Size of Hole Wizard holes
- Sketch relations
- Materials

A Design table can be created in two different ways. The user can start with a blank table and manually insert the parameters, or SOLIDWORKS® can automatically create a design table and then edit the table based on your needs. You can also create the design table externally, outside of SOLIDWORKS®, and then link the Excel file. This method would be most useful for very complex design tables. Creating a design table requires several steps. The best way to learn how to insert and use a design table is to go through the tutorial presented in this chapter.

5.2) CONNECTING ROD CONFIGURATION TUTORIAL

5.2.1) Prerequisites

Before starting this tutorial, you should have completed the following tutorial and be familiar with the following topics.

Prerequisite Tutorials

- Chapter 1 – Connecting Rod Model Tutorial

Prerequisite Topics

- Ability to read dimensions.

5.2.2) What you will learn.

The objective of this tutorial is to introduce you to SOLIDWORKS'® ability to create configurations of a part. You will be creating configurations of the *Connecting Rod* shown in Figure 5.2-1. Specifically, you will be learning the following commands and concepts.

Features

- Add Configuration

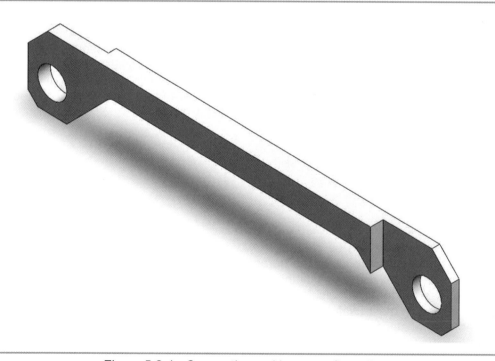

Figure 5.2-1: Connecting rod base configuration

5.2.3) Creating configurations

1) **Open** the **CONNECTING ROD.SLDPRT** file completed in *Chapter 1*.

2) Name the *Default* configuration **Medium**. Do this by entering the *Configuration Manager* 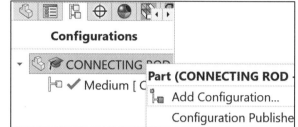 (a tab found on the top of the *Feature Manager Design Tree*) and then slowly double clicking on the default name.

3) **Add** a **Configuration** for the *Connecting Rod* and call it **Short**. Right-click on the part name above the *Medium* configuration and select *Add Configuration*.

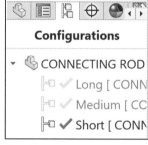

> ➤ See section 5.1 to learn about adding *Configurations*.

4) Add another configuration and name it **Long**.

5) **Double click** on **Short** to make this the active configuration.

6) Get back into the **Feature Manager Design Tree** tab so that the *Short* configuration can be edited.

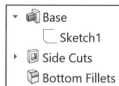

7) **Edit** the **sketch** attached to the **Base** extrude. Do this by expanding *Base* and click on the sketch and select **Edit Sketch**. The dimension values may now be edited by double clicking on them.

8) Change the overall length to **25** and the distance between the holes to **20**. **IMPORTANT!!** Make sure that when you edit the dimensions that you apply it to only **This Configuration**.

9) **Exit** the **Sketch**.

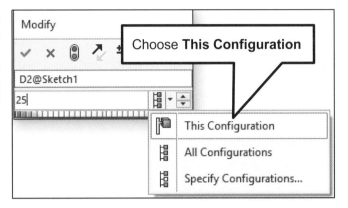

➤ **Problem?** If you don't see the changes that you have made, you should **Rebuild** to see the effect. This icon is usually located at the top.
➤ **Problem?** If you get an error, you will have to look at your original configuration and figure out what is not constrained properly. If your sketch is not completely black, you will most likely have a problem applying configurations.

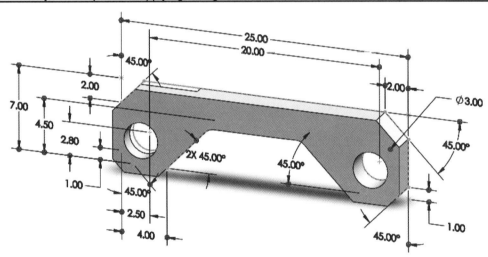

10) Notice that the side cut on the right side has disappeared. Change the dimension of the *Side cuts* as shown. Remember to apply the dimension change to **This Configuration**.

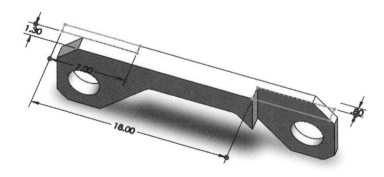

11) Go back to the **Configuration Manager** and switch back and forth between the **Medium** and **Short** configurations to see if it works.

12) Make the **Long** configuration active and then edit the sizes as shown. Change the total length to **55**, the distance between the holes to **50** and the distance to the side cut to **48**.

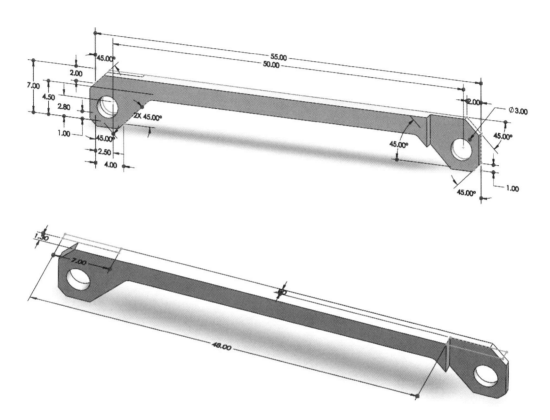

13) Go to the **Configuration Manager** tab and double click on the different configurations and see how the sizes change.

14) **Save**.

5.3) FLANGED COUPLING CONFIGURATION TUTORIAL

5.3.1) Prerequisites

Before starting this tutorial, you should have completed the following tutorial and be familiar with the following topics.

Prerequisite Tutorials

- Chapter 3 – Flanged Coupling Model Tutorial

Prerequisite Topics

- Ability to read dimensions.

5.3.2) What you will learn.

The objective of this tutorial is to introduce you to the SOLIDWORKS'® ability to create configurations of a part. You will be creating configurations of the *Flanged Coupling* shown in Figure 5.3-1. Specifically, you will be learning the following commands and concepts.

Features

- Add Configuration

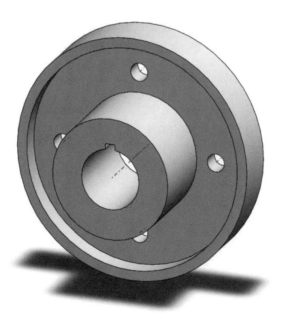

Figure 5.3-1: Flanged Coupling base configuration

5.3.3) Creating Configurations

1) **Open** the **COUPLING.SLDPRT** file created in Chapter 3.

2) **Add** a **Configuration** for the coupling and call it **Right coupling** and rename the Default configuration as **Left coupling**. If a warning window appears, select **YES**. **Double click** on **Right coupling** to make this the active configuration.

3) **Sketch** on the back face of the part.

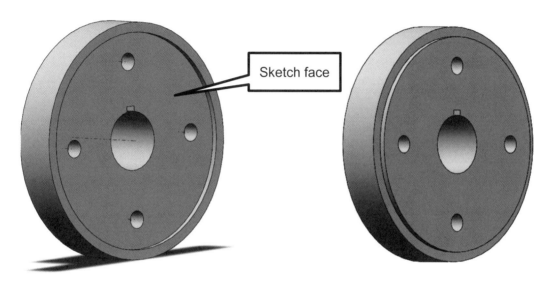

Sketch face

➢ **Problem?** If a weird funnel shape appears and you can't select the face to sketch, this means that you have a filter applied. Press **F5** to view the **Selection Filter** toolbar. Deselect the selected filter.

---

4) **Convert Entities** [Convert Entities] all the features on the sketch face. Convert all the Edges not just the face.

➢ See Chapter 3 for information on *Converting Entities*.

---

5) **Extruded Boss/Base** [Extruded Boss/Base] the sketch **6 mm** away from the face.

6) Go to your *Configuration* tab and click between the **Left** and **Right coupling** to make sure that the configurations work.

7) **IMPORTANT!! Save** your part and **keep** it. It will be used in a later tutorial.

5.4) RIGID COUPLING DESIGN TABLE TUTORIAL

5.4.1) Prerequisites

Before starting this tutorial, you should have completed the following tutorials and be familiar with the following topics.

Prerequisite Tutorials

- Connecting Rod Configuration Tutorial
- Flanged Coupling Configuration Tutorial

Prerequisite Topics

- Ability to read dimensions.

5.4.2) What you will learn.

The objective of this tutorial is to introduce you to the SOLIDWORKS'® ability to create configurations of a part using **Design Tables**. You will be creating configurations of the *Rigid Coupling* shown in Figure 5.4-1. Specifically, you will be learning the following commands and concepts.

Insert

- Design Tables

Figure 5.4-1: Rigid Coupling configurations

5.4.3) Setting up the project

1) **Start SOLIDWORKS** and start a **new part** .

2) Set the drafting standard to **ANSI** and set all the text to **upper case** for notes, tables, and dimensions. (*Options* ⚙ *– Document Properties – Drafting Standard*)

3) Set the units to **MMGS** (millimeters, grams, second) and the **decimal = 0.1**. (*Options* ⚙ *– Document Properties – Units*)

4) Save the part as **RIGID COUPLING.SLDPRT** (**File – Save**). Remember to save often throughout this project.

5.4.4) Creating the base configuration

1) **Sketch** and **Dimension** the following **Circle** on the **Front Plane**.

2) **Extrude** the circle **13 mm**.

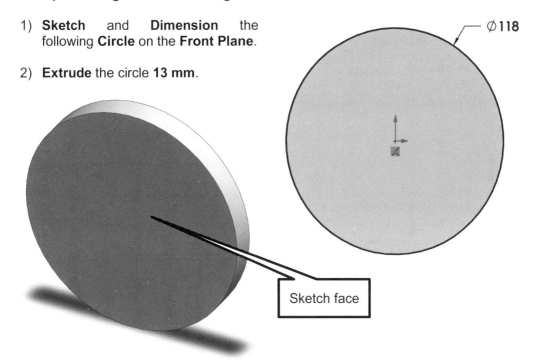

⌀118

Sketch face

3) **Sketch** on the **front face** of the part and then draw and **Dimension** the **Circle** shown.

4) **Extrude** the Circle **13 mm**.

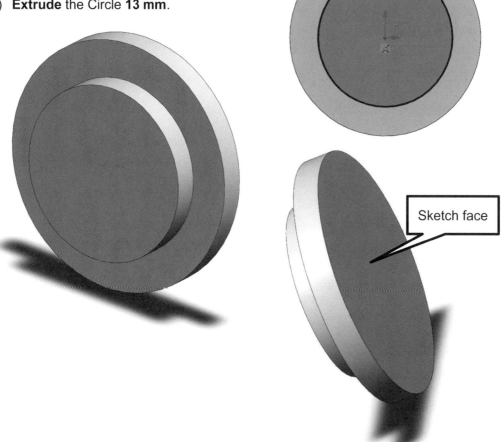

Sketch face

5) **Sketch** on the **back face** of the part and then draw and **Dimension** the **Circle** shown.

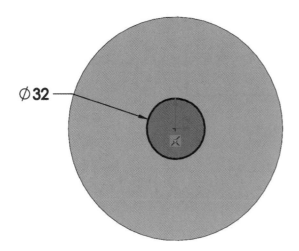

6) **Cut-Extrude** the circle using a **Draft Angle** of **18 degrees**.

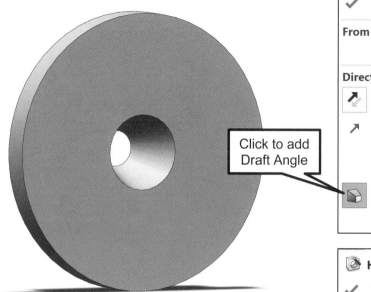

Click to add Draft Angle

7) Add an **8 mm** diameter **Through Hole** 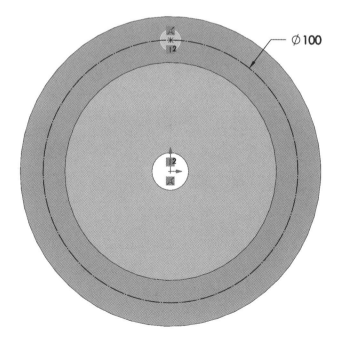 to the base of the coupling. Position the hole as shown in the figure. Use a center of circles to help position the hole. Make this circle a **construction element**.

Ø100

8) Use a **Circular Pattern** 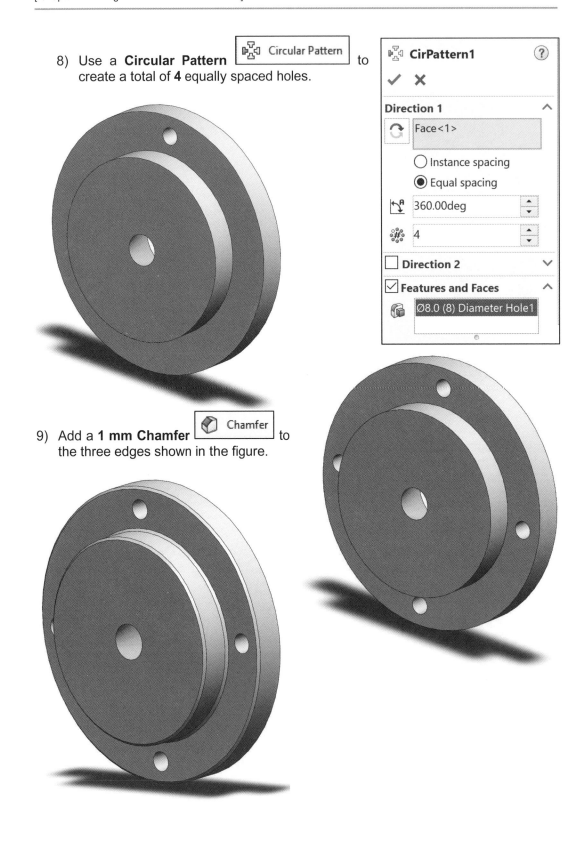 to create a total of **4** equally spaced holes.

9) Add a **1 mm Chamfer** to the three edges shown in the figure.

CirPattern1

Direction 1

Face<1>

○ Instance spacing
◉ Equal spacing

360.00deg

4

☐ **Direction 2**

☑ **Features and Faces**

Ø8.0 (8) Diameter Hole1

5.4.5) Setting up the design table

1) In the *Feature Tree*, right-click on **Annotation** and select **Show Feature Dimensions**. Then pull out all of the dimensions so that they can be seen.

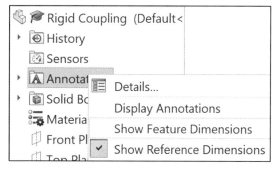

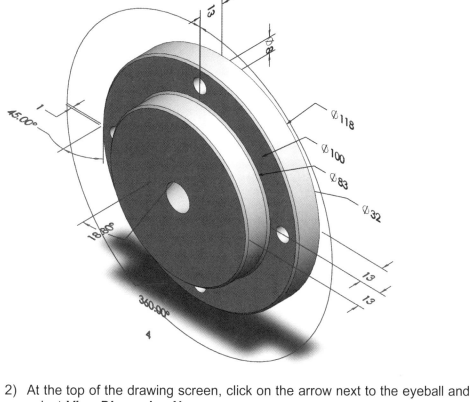

2) At the top of the drawing screen, click on the arrow next to the eyeball and select **View Dimension Names**.

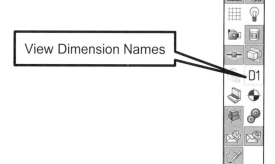

View Dimension Names

3) Name the dimensions. Do this by clicking on a dimension and entering a name in the *Dimension* window. Name the dimensions as shown in the figure.

Enter dimension name here. Don't add "@Sketch1". That will be added automatically.

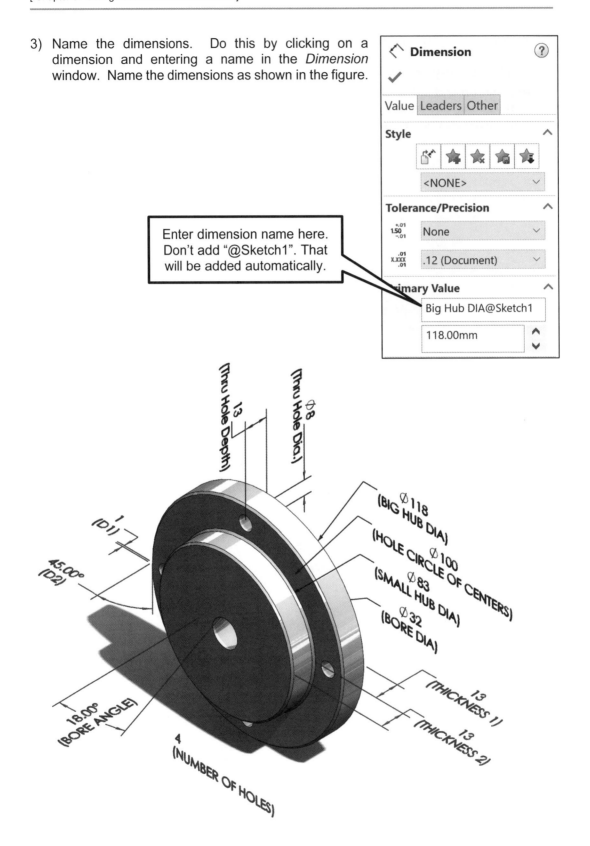

4) Insert the Design Table by selecting from the top pull-down menu, ***Insert – Tables – Excel Design Table…*** Have SOLIDWORKS® create the table automatically and allow for edits. A *Dimensions* window will appear; select the dimensions that are to be included in the design table and select **OK**. It may not detect all of the dimensions in the model. We will have to add those later. If you get a warning window, select **OK**.

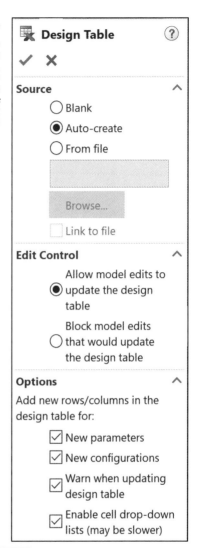

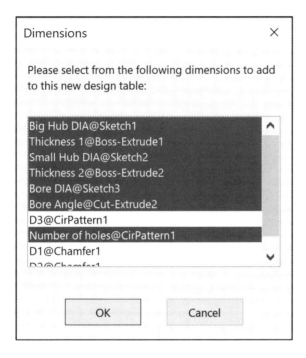

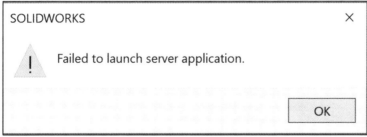

5) Your design table should look something like what is shown. We need to manually add any dimensions that were not added. When you are typing, pay attention to all **spaces**, **punctuations**, and **capitals**. The second table shows the completed table with all the dimensions. There should be a total of 9 dimensions that we want to control. When you are done, **close** the table by clicking in the modeling window.

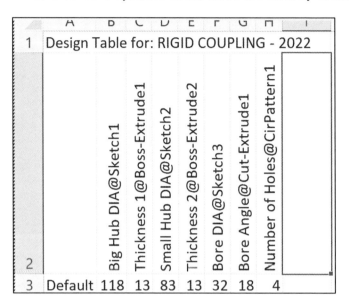

	A	B	C	D	E	F	G	H	I
1	Design Table for: RIGID COUPLING - 2022								
2		Big Hub DIA@Sketch1	Thickness 1@Boss-Extrude1	Small Hub DIA@Sketch2	Thickness 2@Boss-Extrude2	Bore DIA@Sketch3	Bore Angle@Cut-Extrude1	Number of Holes@CirPattern1	
3	Default	118	13	83	13	32	18	4	

> **NOTE:** Your "@Sketch", "@Boss-Extrude", and others may have different numbers depending on how your part was constructed. If you get errors, this is the first thing to check.

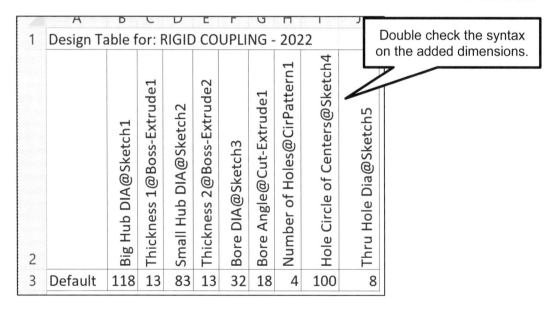

	A	B	C	D	E	F	G	H	I	J
1	Design Table for: RIGID COUPLING - 2022									
2		Big Hub DIA@Sketch1	Thickness 1@Boss-Extrude1	Small Hub DIA@Sketch2	Thickness 2@Boss-Extrude2	Bore DIA@Sketch3	Bore Angle@Cut-Extrude1	Number of Holes@CirPattern1	Hole Circle of Centers@Sketch4	Thru Hole Dia@Sketch5
3	Default	118	13	83	13	32	18	4	100	8

Double check the syntax on the added dimensions.

6) If you get an error, click on the **Configuration Manager**, expand *Tables*, right-click on the **Design Table**, and Select **Edit Table**. A *Warning* window and an *Add Rows and Columns* window will appear. Select **OK**. All of the correct dimensions will be highlighted in pink on your model. If they are not pink, that means something is wrong. It is most

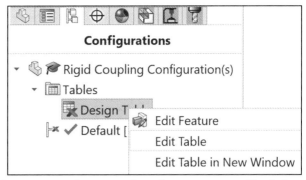

likely a syntax error. Try to fix the error and repeat the above steps until all the dimension that we are including in the design table are pink.

7) Edit your table to rename the original configuration and then add two new configurations. When you're done, close your table. A window will appear saying that you are about to generate configurations; select **OK**.

Design Table for: RIGID COUPLING - 2022	Big Hub DIA@Sketch1	Thickness 1@Boss-Extrude1	Small Hub DIA@Sketch2	Thickness 2@Boss-Extrude2	Bore DIA@Sketch3	Bore Angle@Cut-Extrude1	Number of Holes@CirPattern1	Hole Circle of Centers@Sketch4	Thru Hole DIA@Sketch5
Small	118	13	83	13	32	18	4	100	8
Medium	216	38	145	38	75	20	6	180	10
Large	330	57	230	57	110	22	8	280	12

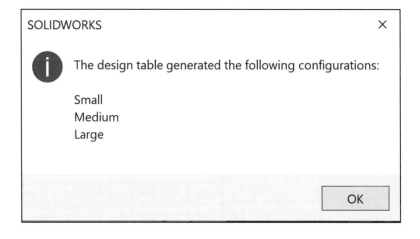

SOLIDWORKS   ✕

ℹ The design table generated the following configurations:

Small
Medium
Large

OK

8) Click on the **Configuration Manager**. Notice that there are now three configurations. Click between them to see the changes.

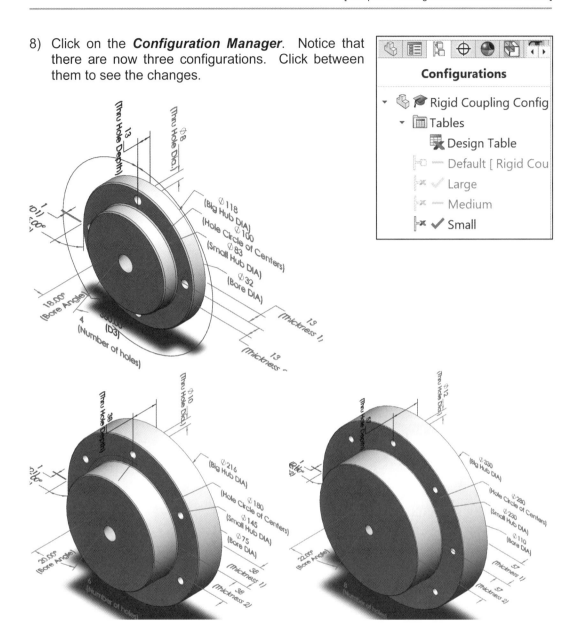

5.4.6) Suppressing features and changing material

1) **Edit** the design **table**. Enter **$state@Chamfer1** into the next empty cell. The **$state** variable allows us to Suppress and Unsuppress a feature.

Design Table for: RIGID COUPLING - 2022									
	Big Hub DIA@Sketch1	Thickness 1@Boss-Extrude1	Small Hub DIA@Sketch2	Thickness 2@Boss-Extrude2	Bore DIA@Sketch3	Bore Angle@Cut-Extrude1	Number of Holes@CirPattern1	Hole Circle of Centers@Sketch4	Thru Hole Dia@Sketch5	$state@Chamfer1

2) **Exit** the table and re-**edit** the **table**. Notice that the table has auto populated. Change the *Small* configuration from **U** to **S**.

Design Table for: RIGID COUPLING - 2022										
	Big Hub DIA@Sketch1	Thickness 1@Boss-Extrude1	Small Hub DIA@Sketch2	Thickness 2@Boss-Extrude2	Bore DIA@Sketch3	Bore Angle@Cut-Extrude1	Number of Holes@CirPattern1	Hole Circle of Centers@Sketch4	Thru Hole Dia@Sketch5	$state@Chamfer1
Small	118	13	83	13	32	18	4	100	8	S
Medium	216	38	145	38	75	20	6	180	10	U
Large	330	57	230	57	110	22	8	280	12	U

3) **Exit** the table and change between the configurations. Notice that the chamfers disappear on the *Small* configuration.

4) Edit the table and enter **$Library:material@***File Name* in to the next empty cell. Make the *Small* configuration **Oak** and the *Large* configurations **Brass** as shown in the figure. The Medium configuration cell is empty because we are leaving that one as modeled.

Design Table for: RIGID COUPLING - 2022	Big Hub DIA@Sketch1	Thickness 1@Boss-Extrude1	Small Hub DIA@Sketch2	Thickness 2@Boss-Extrude2	Bore DIA@Sketch3	Bore Angle@Cut-Extrude1	Number of Holes@CirPattern1	Hole Circle of Centers@Sketch4	Thru Hole Dia@Sketch5	$state@Chamfer1	$Library:material@RIGID COUPLING - 2022
Small	118	13	83	13	32	18	4	100	8	S	SOLIDWORKS Materials:Oak
Medium	216	38	145	38	75	20	6	180	10	U	
Large	330	57	230	57	110	22	8	280	12	U	SOLIDWORKS Materials:Brass

5) **Exit** the table and change between the configurations. Notice that the configurations change material.

6) **Save** your part.

NOTES:

CONFIGURATIONS IN SOLIDWORKS® PROBLEMS

P5-1) Create a configuration of the part given in P1-1. Make the new configuration so that every dimension except for the thickness is twice as big as original part and determine the new configuration's weight.

P5-2) Create a configuration of the part given in P1-2. Make the new configuration so that the 20 mm diameter cylinder changes to 40 mm and the length doubles. Determine the new configuration's weight.

P5-3) Create a configuration of the part given in P1-3. Create a configuration where the base thickness is 1.00 inch (instead of 0.50 inch).

P5-4) Create a configuration of the part given in P1-4. Create a configuration where the thickness of the two upright walls is 0.25 inch (instead of 0.50 inch).

P5-5) Create a configuration of the part given in P1-5. Make the added configuration twice as long as the original part and determine the new configuration's weight.

P5-6) Create a configuration of the part given in P1-6. Create a configuration where the solid cylinder on top is half the height/long and the material is Pine.

P5-7) Create a configuration of the part given in P1-7. Create a configuration where the holes are suppressed, and the part material is Brass.

P5-8) Create a configuration of the part given in P1-8. Make the added configuration have an increased diameter. Make the n1.50 diameter, n3.00 and determine the new configuration's weight.

P5-9) Create a configuration of the part given in P1-9. Create a configuration where the hole is suppressed, and the thickness of the part is doubled. This includes the thickness of the tab.

P5-10) Create a configuration of the part given in P1-10. Create a configuration where the hole is located 1.50 inches high (instead of 1.25) and the height of the part is 2.5 inches.

P5-11) Create a configuration of the part given in P1-11. Create a configuration where the size of the large hole is changed to 25 mm (instead of 50 mm) and the height at which the angle face starts changes to 25 mm (instead of 50 mm).

P5-12) Create a configuration of the part given in P1-12. Create a configuration where the two larger cylinder diameters are doubled.

P5-13) Create a configuration of the part given in P1-13. Create a configuration where the 100 mm dimension changes to 150 mm.

P5-14) Create a configuration of the part given in P1-14. Create a configuration where the larger hole is half the size, and the smaller holes are eliminated.

P5-15) Create a configuration of the part given in P3-1. Suppress the holes in the new configuration and determine its weight.

P5-16) Create a design table of the part given in P3-5. The *Small* configuration is the original model.

	A	B	C	D	E	F
1	Design Table for: P3-5					
2		Bore DIA@Sketch1	Hub DIA@Sketch1	Inner DIA@Sketch1	Root DIA@Sketch1	Crest DIA@Sketch1
3	Small	50	65	80	133	138
4	Medium	60	75	90	145	153
5	Large	70	85	100	158	166

P5-17) Create a design table of the part given in P3-6.

	Big Hub DIA@Sketch1	Small Hub DIA@Sketch1	Bore DIA@Sketch1	Spoke Radius@Sketch2	Number of Spokes@CirPattern1	$state@Cut-Extrude1		$Library:material@p5-17
Small	80	70	42	80	4	U	SOLIDWORKS Materials:PE High Density	
Large	160	140	84	160	8	S	SOLIDWORKS Materials:AISI 1020	

CHAPTER 6

STATIC FEA IN SOLIDWORKS®

CHAPTER OUTLINE

6.1) WHAT IS FEA? 2

6.2) THEORY BACKGROUND 2

    6.2.1) Engineering Stress and Strain 2

6.3) ADD-INS 4

6.4) SIMULATION 5

    6.4.1) Simulation Process 5

    6.4.2) Simulation Study Tree 6

    6.4.3) Static Simulation 7

    6.4.4) Generating Reports 8

6.5) CONNECTING ROD STATIC FEA TUTORIAL 9

    6.5.1) Prerequisites 9

    6.5.2) What you will learn 9

    6.5.3) Setting up the static analysis 11

    6.5.4) Running the analysis 14

STATIC FEA IN SOLIDWORKS PROBLEMS 21

CHAPTER SUMMARY

In this chapter, you will learn how to perform a static FEA in SOLIDWORKS®. You will apply forces and torques to an existing model, and determine the resulting stress, strain, and displacement. By the end of this chapter, you will be able to run static FEAs and interpret the results.

6.1) WHAT IS FEA?

Finite element analysis (FEA) or **finite element method** (FEM) is a method that allows you to obtain resulting stresses, strains, heat flow, fluid flow, and other quantities for a complex object under load. These quantities are usually governed by unmanageable mathematical equations. How does FEM/FEA do this? FEM/FEA beaks up the part into many small pieces of simple shapes called **elements**. This allows the software to analyze a very complex system by analyzing many very small simple elements. However, with simplicity comes the loss of accuracy. It is a trade-off. The small pieces used to simplify the analysis are created by a process called **meshing**. The size and shape of the mesh or pieces can be controlled to help minimize error or computing time depending on the priority. Once the part is analyzed, a map (usually a color code) is created to allow a visual view of the results.

6.2) THEORY BACKGROUND

6.2.1) Engineering Stress and Strain

Consider a rod being pulled at both ends by a force as shown in Figure 6.2-1. As the force pulls on the rod, it stretches (maybe by only an infinitesimal amount.) Even if the stretch or the deformation is not visible, this force still causes the rod's material to experience what is called *stress* (σ). The deformation of the material can be quantified by a term called strain (ε).

Engineering stress is equal to the pulling force divided by the original cross-sectional area of the material. Stress is measured in Pascal's (Pa) which is equivalent to Newton's per meter squared (N/m^2). Stress can also be measured in Kilo-pounds per square inch (ksi). **Engineering strain** is equal to the change in the length of the material over its original length. Strain is essentially unitless because it is length over length, but it is usually written as (mm/mm) or (in/in).

When a material is stressed, the bonds between the material's atoms are being pulled. If the force is not large enough to break these bonds, the material will experience elastic deformation. If the force is increased and becomes large enough, these bonds will break, and the material will then experience permanent or plastic deformation.

A part will plastically deform at locations where the stress exceeds the yield strength of the material. The **yield strength** is the maximum stress that a material can sustain before plastic or permanent deformation occurs. Think of a paper clip. If a large enough force is applied to a paper clip, it will bend and remain bent. This is plastic deformation. In most cases, plastic deformation is bad and should be avoided. The **ultimate** or **tensile strength** is the maximum stress that a material can sustain before fracture.

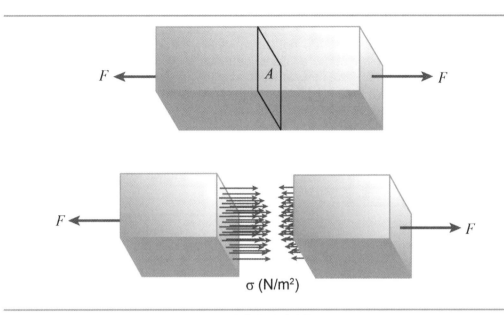

Figure 6.2-1: Axial Stress

There are many types of stress and strain. What is described above is called axial stress and strain. However, there is bending and shear stress and strain as well. The basic idea behind them all is that you want the actual stress on a part to be low enough so that it does not exceed the yield strength of the material and the part does not permanently deform.

These YouTube® videos may also help clarify these concepts.
- Elastic and Plastic Deformation (https://youtu.be/D3kzv2s8z5g)
- Tensile testing (https://youtu.be/4Q_-O-LnSGY)
- Modes of Fracture (https://youtu.be/Beqe-adBy_U)

6.3) ADD-INS

Add-Ins are features that are not automatically loaded when SOLIDWORKS® starts. The type of *Add-Ins* that are available depends on your license. Generally, you don't want to load these *Add-Ins* all the time because they will slow down the speed of the program. The *Add-Ins* are accessed through the pull-down menu (**Tools – Add-Ins…**). The *Add-Ins* option window is shown in Figure 6.3-1.

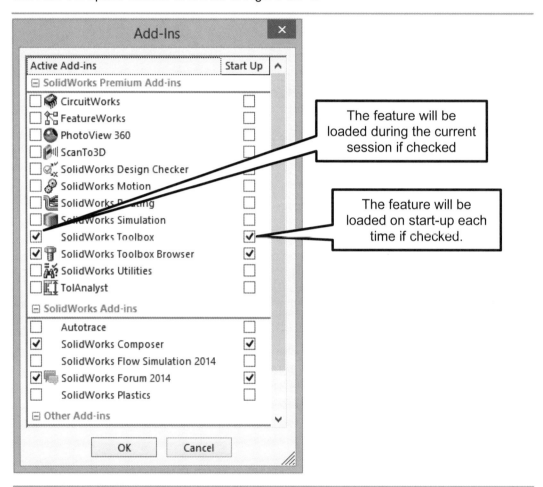

Figure 6.3-1: Add-Ins option window

6.4) SIMULATION

6.4.1) Simulation Process

SOLIDWORKS® *Simulation* is designed to analyze parts that are under the influence of forces, heat, or impact as well as other types of loads. This analysis is done through a process called the **finite element method** (FEM) or **finite element analysis** (FEA). FEM/FEA breaks up the part into many small pieces of simple shapes called **elements**. This allows the software to analyze a very complex system more simply. However, with simplicity comes loss of accuracy. It is a trade-off. The small pieces used to simplify the analysis are created by a process called **meshing**. The size and shape of the mesh or pieces can be controlled. Once the part is analyzed, a map (usually a color code) is created to allow a visual view of the results. Figure 6.4-1 shows a part model, that same model meshed, and the results of an FEA performed on the part. The types of simulation studies that can be run are shown in the *Study* window shown in Figure 6.4-2. To start a simulation, use the following steps.

Starting a simulation

1) Pull-down menu: **Tools – Add-Ins…**
2) Select **SOLIDWORKS Simulation** to be added.
3) Select the **Simulation tab**.

4) Select **New Study**.
5) Name the study.
6) Select **Static** or another type of study.

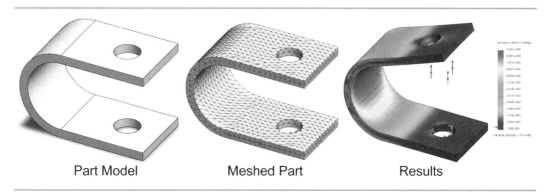

| Part Model | Meshed Part | Results |

Figure 6.4-1: FEA example

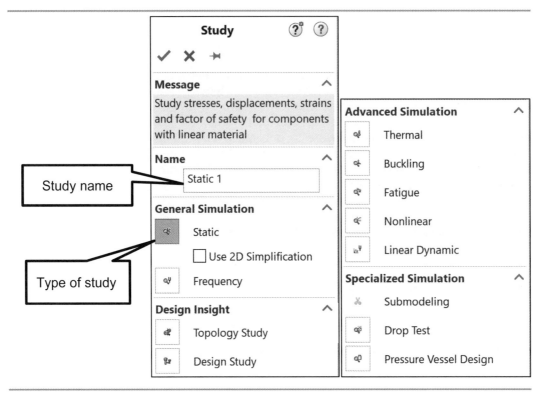

Figure 6.4-2: Study window

6.4.2) Simulation Study Tree

After a new simulation study is started, a *Simulation Study Tree* will appear below the *Feature Manager Design Tree*. The *Study Tree* essentially lists the options that need to be specified before the FEA may be run. These options are usually applied in order from top to bottom. This is also where the results of the analysis can be viewed. Each type of simulation will have a different *Simulation Study Tree*. Figure 6.4-3 shows a few examples of different *Simulation Study Trees*.

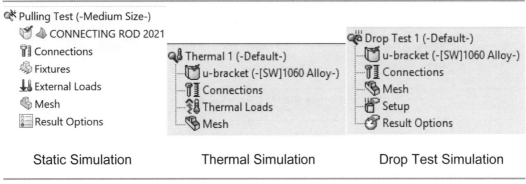

Figure 6.4-3: Simulation Study Tree examples

6.4.3) Static Simulation

Static Simulation 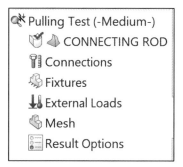 analyzes a part under forces, torques, and pressures. It can calculate displacements, reaction forces, strains, stresses, and factor of safety distribution. There are several options in the *Simulation Study Tree* that need to be specified. Reading from top to bottom, there are:

- <u>Connections:</u> This is where you would specify any contacts, springs, bolts, bearings, or welds.
- <u>Fixtures:</u> This is where you specify any supports such as fixed geometry or an elastic support.
- <u>External Loads:</u> You may apply a force or a torque to your part. Figure 6.4-4 shows the force/torque option window.
- <u>Mesh:</u> You can specify your mesh size and shape.

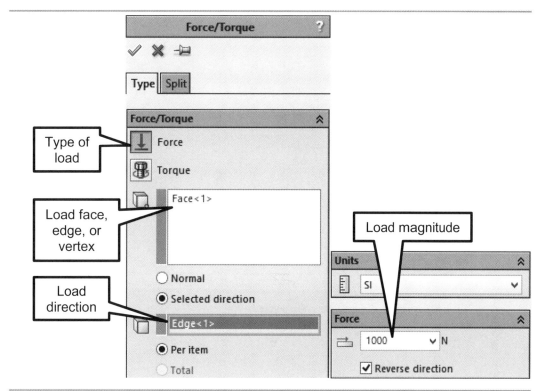

Figure 6.4-4: Force and Torque option window

6.4.4) Generating Reports

A **report** 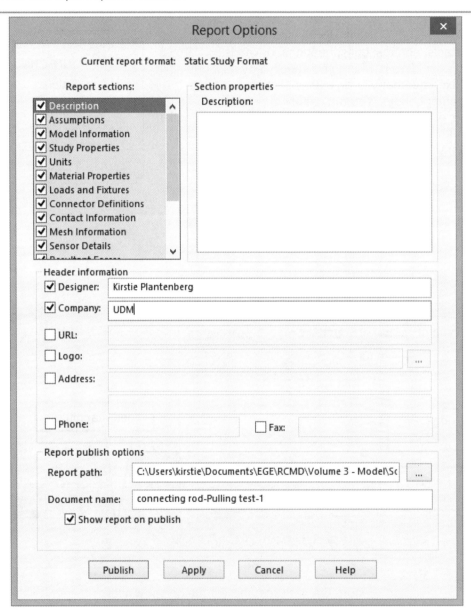 is an MS Word® document which gives pertinent information about your analysis. It is nice to use if a report is being written for class or business. Figure 6.4-5 shows the *Report* options window.

Figure 6.4-5: Report options window

6.5) CONNECTING ROD STATIC FEA TUTORIAL

6.5.1) Prerequisites

To complete this tutorial, you should have completed the listed tutorial. It would also be helpful if you were familiar with the listed topics.

- Chapter 1 – Connecting Rod Part Model Tutorial
- Passing familiarity with the concepts of force, stress, and strain.

6.5.2) What you will learn

The objective of this tutorial is to introduce you to the SOLIDWORKS®' static simulation capabilities. In this tutorial you will be analyzing the *Connecting Rod* model under load (modeled in Chapter 1). The part print of the *Connecting Rod* is shown in Figure 6.5-1. Specifically, you will be learning the following commands and concepts.

Simulation

- Add-Ins
- New study
- Static simulation
- Simulation study tree
- Stress
- Strain
- Generating a report

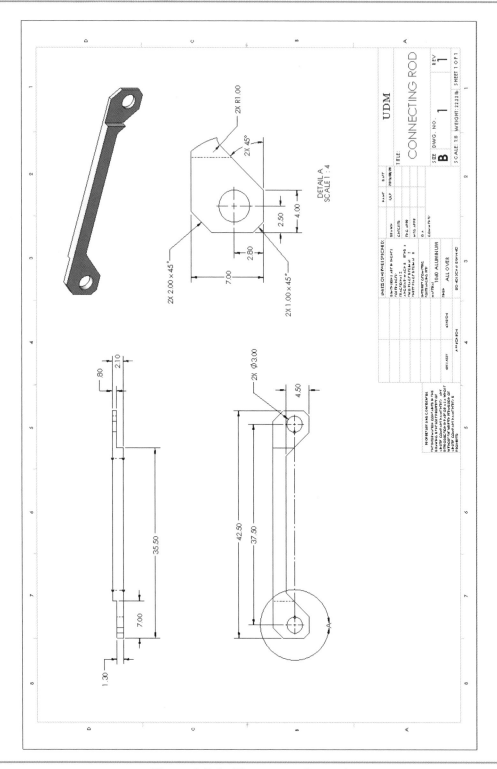

Figure 6.5-1: The connecting rod

6.5.3) Setting up the static analysis

1) Open **CONNECTING ROD.SLDPRT** (modeled in Chapter 1). If you have completed the configuration tutorial, activate the **Medium Size** configuration. Remember to **Save** often throughout this tutorial.

2) Load the **SOLIDWORKS Simulation** *Add-In* by selecting the **SOLIDWORKS Add-Ins**

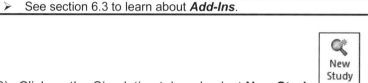

tab and then activate the **SOLIDWORKS Simulation** Add-in. Notice that a *Simulation* tab appears in the *Command Manager*. This may take awhile.

> ➤ See section 6.3 to learn about **Add-Ins**.

3) Click on the *Simulation tab* and select **New Study**.

4) Create a new **Static Study** and name it **Pulling test**.

> ➤ See section 6.4 to learn about **Simulation**.

Study

Message

Study stresses, displacements, strains and factor of safety for components with linear material

Name

Pulling Test

General Simulation

Static

5) A *Simulation Study Tree* should have appeared below the *Feature Manager Design Tree*. The steps need to be completed from top to bottom. The first step is to assign a **material** to the part. This should have already been done since we assigned the material in our model. Make sure that next to the part name it lists the part material.

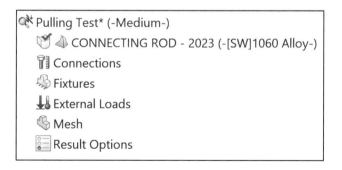

Pulling Test* (-Medium-)
CONNECTING ROD - 2023 (-[SW]1060 Alloy-)
Connections
Fixtures
External Loads
Mesh
Result Options

> ➤ See section 6.4.1 to learn about the **Simulation Study Tree**.

6) Fix the left hole of the connecting rod.
   a) Right-click on **Fixtures** and select **Fixed Geometry.**
   b) In the *Fixture* window, activate the **Fixed Geometry** option.
   c) Select the inside surface of the left hole. Green arrows should appear indicating which surface(s) are fixed.
   d)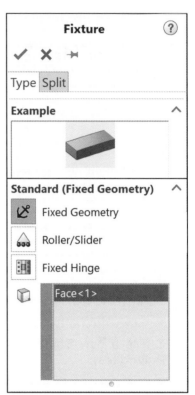

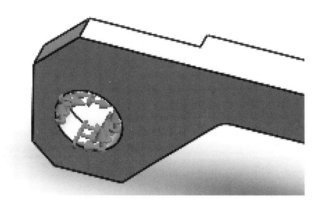

7) Apply a load to the free end of the *Connecting Rod*.
   a) In the *Simulation Study Tree*, right-click on **External Loads** and select **Force.**
   b) Select the inner surface of the right-side hole as the surface where the force will be applied.
   c) Click on the **Selected direction** radio button and use one of the long edges of the part to define the direction of pull.   You may have to **Reverse direction** so that the part is pulled and not compressed.
   d) Enter a force of **1000 N**.
   e) ☑

➢ See section 6.4.3 to learn about *Static Simulation*.

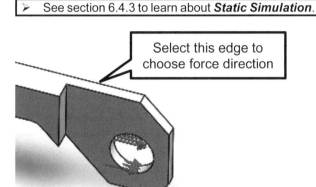

Select this edge to choose force direction

8) Mesh the part. In the *Simulation Study Tree*, right-click on **Mesh** and then select **Create Mesh**. Use the default mesh parameters.

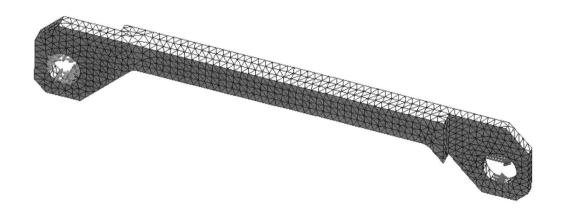

6.5.4) Running the analysis

1) **Run this Study** [Run This Study]. The analysis may take a while to run. The analysis time will depend on your computer speed. After the analysis is completed, the part will be mapped in colors. A legend indicating what the colors mean will also appear. The default legend will indicate stress. Note that the highest stress is below the material's yield strength. The maximum stress should be around 4.6e+006 N/m^2 (i.e., 4.6×10^6 Pa) and the yield strength is 2.757e+007 N/m^2. This means that the part will not break. You may double click on the other result plots to see the part's **Displacement** and **Strain**.

Pulling Test (-Medium Size-)
  CONNECTING ROD 2021 (-[SW]1060 Alloy-)
  Connections
  Fixtures
    Fixed-1
  External Loads
    Force-1 (:Per item: -1,000 N:)
  Mesh
    Mesh Quality Plot
      Quality1 (-Mesh-)
  Result Options
  Results
    Stress1 (-vonMises-)
    Displacement1 (-Res disp-)
    Strain1 (-Equivalent-)

> ➢ See section 6.2.1 to learn about **Stress & Strain.**

von Mises (N/m^2)

4.605e+06
4.145e+06
3.684e+06
3.224e+06
2.763e+06
2.303e+06
1.842e+06
1.382e+06
9.215e+05
4.610e+05
5.155e+02

---▶ Yield strength: 2.757e+07

2) Increase the applied force to **10,000 N**.
    a) Right-clicking on **Force – 1** and selecting **Edit Definition**.
    b) Change the force to **10,000 N**.
    c) ✓
    d) Re-**Run** the analysis.
    e) **Notice** the red arrow on the stress key. It indicates the yield strength of the material. Since the yield strength is within the values of stress experienced by the rod, there is a good chance the part will fail.

von Mises (N/m^2)

4.667e+07
4.200e+07
3.733e+07
3.267e+07
2.800e+07
2.334e+07
1.867e+07
1.400e+07
9.337e+06
4.670e+06
4.286e+03

▶ Yield strength: 2.757e+07

3) Change the force magnitude to **550 N** and the force direction as indicated in the figure. Re-**Run** the analysis.

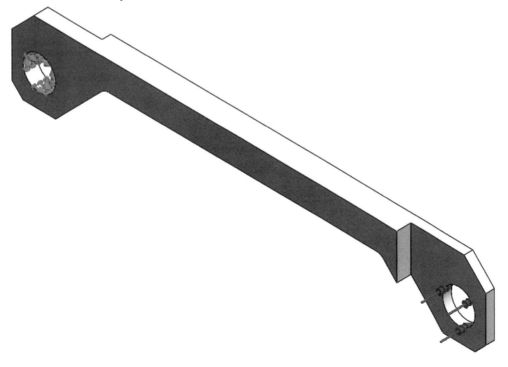

4) Look at the results. Does the part fail on the bases of stress? Don't worry if the stress values are slightly different.

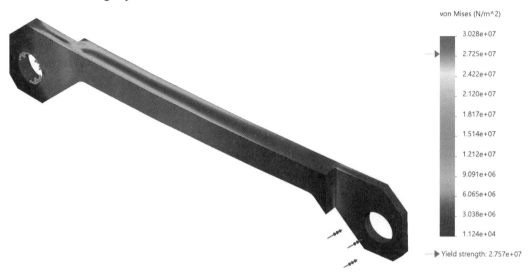

von Mises (N/m^2)

3.028e+07
2.725e+07
2.422e+07
2.120e+07
1.817e+07
1.514e+07
1.212e+07
9.091e+06
6.065e+06
3.038e+06
1.124e+04

Yield strength: 2.757e+07

5) Locate the maximum stress on the part.
   a) Under *Results*, right-click on **Stress**.
   b) Select **Chart Options**.
   c) Activate the **Show max annotation**.
   d) **Note** that the maximum stress values appear on the part. The maximum stress is indeed greater than the material's yield strength. Under these conditions, this part has a high probability of not performing as intended. Also note that the high stress occurs at a sharp corner or an area of high-stress **concentration**. Any time an area changes shape or direction abruptly, stress is amplified.

6) **Save** the part as **CONNECTING ROD – CHANGES.SLDPART.** Click on the *Model* tab (at the bottom) so that the model can be edited. Add a **0.75-inch Fillet** to both inside corners.

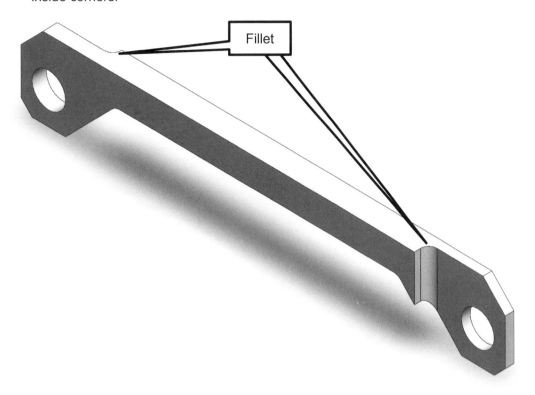

7) Re-enter the *Simulation* tab (at the bottom), re-**mesh** and re-**run** the analysis. Notice that the maximum stress decreased but did not fall below the yield strength of the material.

8) Let's adjust the fillet to make it have a gentler slope to see if we can reduce the stress even more.
    a) Re-enter the *Model* tab.
    b) **Edit** the **Fillet**.
    c) Change the Fillet Parameter from *Symmetric* to **Asymmetric**.
    d) We want to make the edge of the fillet that is touching the surface with the hole to have a larger radius. Change it from 0.75 to **1.25 inch**.

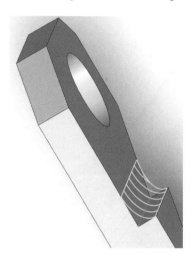

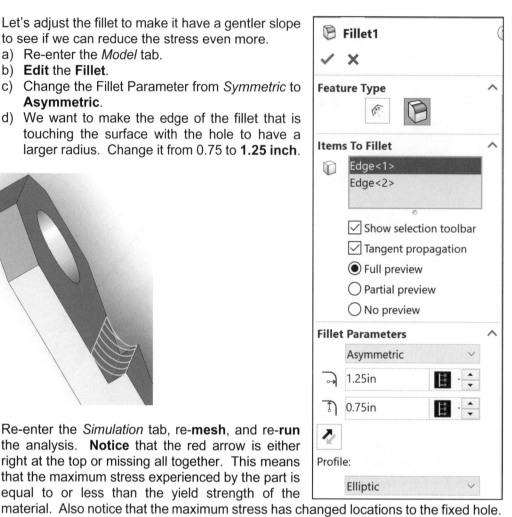

9) Re-enter the *Simulation* tab, re-**mesh**, and re-**run** the analysis. **Notice** that the red arrow is either right at the top or missing all together. This means that the maximum stress experienced by the part is equal to or less than the yield strength of the material. Also notice that the maximum stress has changed locations to the fixed hole.

10) Do you think the final maximum stress is acceptable or would you try to make more changes? **Save** your part as **CONNECTING ROD – CHANGES 2.SLDPART**. On your own, make some changes that you think would reduce the maximum stress even further.

11) What we just did is called **design iteration**. Slowly improving the design of an object based on the results of the FEA. It is always a good idea to save your design iterations under different file names so that you can go back to previous iterations if the next step doesn't work like you planned.

12) Publish a **Report**  of this analysis. Then, look at the report and see what information is contained in it. Note that a report cannot be generated unless you have MS Word® installed on your computer.
   a) Include all report sections.
   b) Designer = **Your name**
   c) Company = **Your school/company name**
   d) Select a file location
   e) Name your document = **Your name + connecting rod**

> ➢ See section 6.4.4 to learn about **Generating a Report**.

NOTES:

STATIC FEA IN SOLIDWORKS PROBLEMS

P6-1) Use SOLIDWORKS® to create a solid model of the following 1345 Aluminum part. Using SOLIDWORKS® Simulation, determine the maximum force (to the nearest 1/2 lb.) that can be applied to the part so that the maximum displacement is less than 0.2 mm. The bottom hole is fixed, and the load is applied upward to the inside of the top hole. Generate a report. Include the following information at the end of your report. Dimensions are given in inches.

1. Maximum force = _____
2. Maximum displacement = _____
3. Material's Yield strength = _____
4. Maximum stress experienced by the part = _____
5. A picture of where the maximum stress is located.

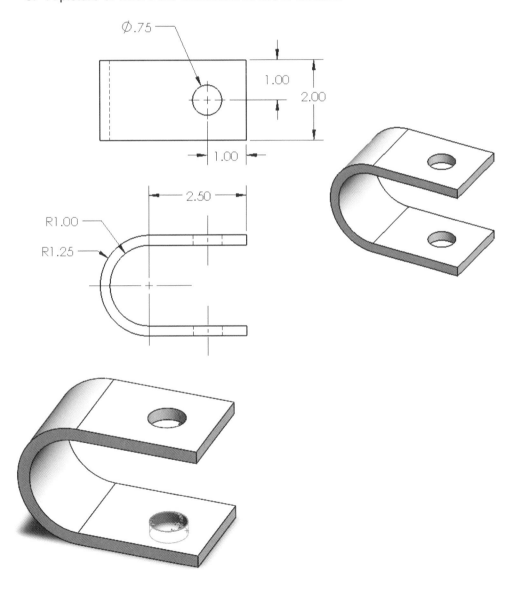

P6-2) Use SOLIDWORKS® to create a solid model of the following Acrylic part. Using SOLIDWORKS® Simulation, apply a downward force of 5 Newtons to the right end of the part as shown in the Front view. The left end of the part is fixed. Generate a report. Include the following information at the end of your report. Dimensions are given in inches. Note that TYP means typical. So, the thickness of the part is 0.25 all over.

1. Material's yield strength = _____
2. Maximum stress experienced by the part = _____
3. Is there a possibility of the part failing due to stress?
4. Include a picture of where the maximum stress is located.
5. If the design parameters state that the part cannot deflect more than 3 mm, does the part perform as intended?

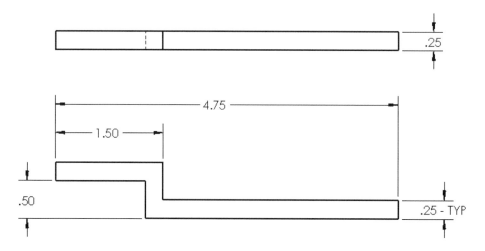

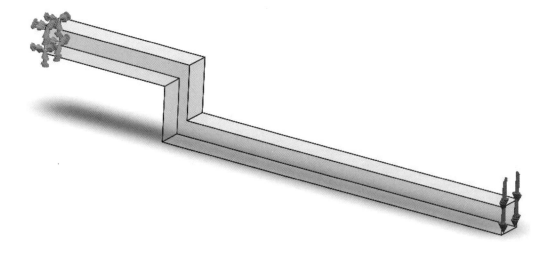

P6-3) Use SOLIDWORKS® to create a solid model of the following AISI 1020 Steel Cold Rolled part. Using SOLIDWORKS® Simulation, determine the maximum torque that can be applied to the part before it yields. The left end of the part is fixed, and the torque is applied to the right end. Generate a report. Include the following information at the end of the report. Dimensions are given in millimeters.

1. Maximum safe torque = _____ N-m
2. Material's yield strength = _____ Pa
3. Maximum stress experienced by the part = _____ Pa
4. A picture of where the maximum stress is located.

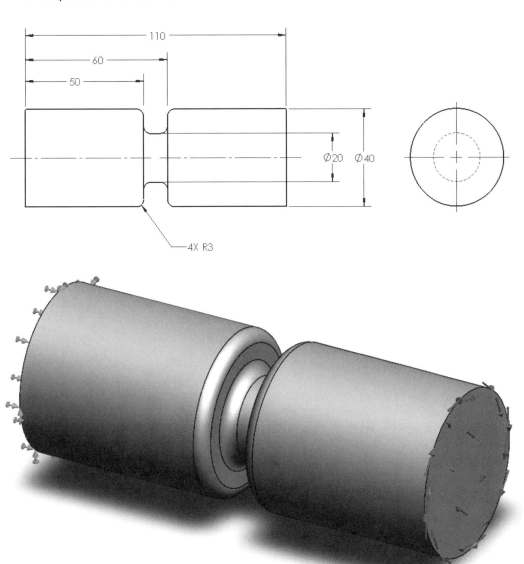

P6-4) Using SOLIDWORKS® Simulation, determine the maximum downward force that can be applied to the *Medium* configuration of the *Connecting Rod* modeled in Chapter 1 before it yields. The left hole of the part is fixed, and the force is applied to the right hole. Generate a report. Include the following information at the bottom of the report.

1. Maximum safe force = _____
2. Material's yield strength = _____
3. Maximum stress experienced by the part = _____
4. A picture of where the maximum stress is located.

Change the fillet radius to 2.00 in.

Re-mesh and re-run the simulation using the force you found in part 1. Indicate the following information in your report.

5. Maximum stress experienced by the part with a 2.00 in fillet = _____
6. The reason that the stress decreased when the fillet radius increased is because _____.

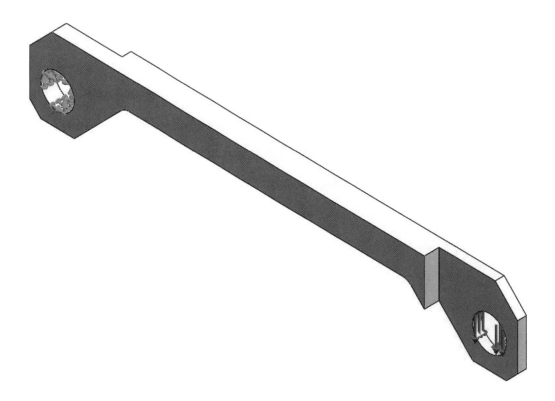

P6-5) Use SOLIDWORKS® to create a solid model of the following Oak ruler. Using SOLIDWORKS® Simulation, determine the maximum downward force that can be applied to the end of the ruler before it yields or breaks. The left end of the ruler is fixed, and the force is applied to the right end. Generate a report. Include the following information at the bottom of the report. Dimensions are in inches.

1. Maximum safe force = _____
2. Material's yield strength = _____
3. Maximum stress experienced by the part = _____
4. A picture of where the maximum stress is located.

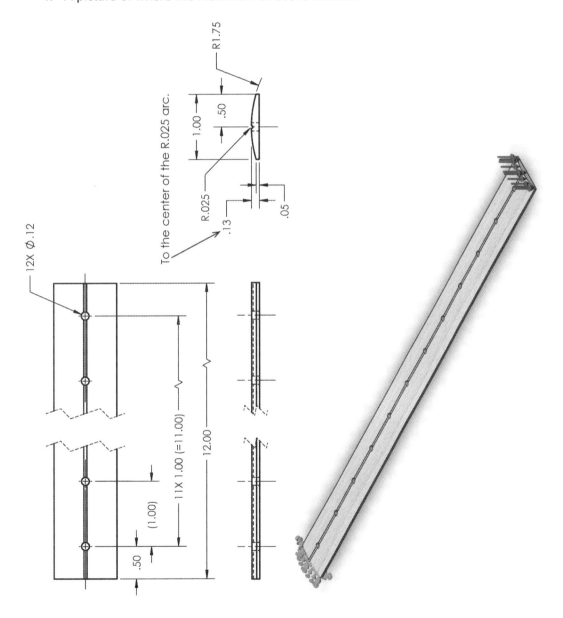

NOTES:

CHAPTER 7

BASIC ASSEMBLIES IN SOLIDWORKS®

CHAPTER OUTLINE

7.1) ASSEMBLIES IN SOLIDWORKS 2

    7.1.1) New Assembly 2

    7.1.2) Inserting Parts 3

    7.1.3) Feature Manager Design Tree 4

    7.1.4) Copying Parts 5

7.2) MATES 6

    7.2.1) The Mate command 6

    7.2.2) Standard Mates 8

7.3) TOOLBOX 8

    7.3.1) Adding the Toolbox 8

    7.3.2) Inserting a Toolbox Component 8

7.4) FLANGED COUPLING ASSEMBLY TUTORIAL 11

    7.4.1) Prerequisites 11

    7.4.2) What you will learn 11

    7.4.3) Shaft 13

    7.4.4) Key 14

    7.4.5) Setting up the assembly 14

    7.4.6) Creating mates 17

    7.4.7) The Toolbox 22

    7.4.8) Exploded assembly 24

BASIC ASSEMBLIES IN SOLIDWORKS® PROBLEMS 27

CHAPTER SUMMARY

In this chapter, you will learn how to create assemblies in SOLIDWORKS®. An assembly is the joining together of two or more parts using constraints or mates. SOLIDWORKS® has many different types of mates that can be used to assemble parts. This chapter will focus on standard mates. You will also learn how to retrieve components from the toolbox library. Components such as bolts and nuts. By the end of this chapter, you will be able to create assemblies with standard mates and toolbox components.

7.1) ASSEMBLIES IN SOLIDWORKS

An assembly consists of several parts that are joined together to perform a specific function (e.g., a bicycle). The assembly may be disassembled without destroying any part of the assembly. SOLIDWORKS® makes it very easy to assemble parts using *Mates*. *Mates* will be talked about in the next section.

7.1.1) New Assembly

You start a new assembly when you want to assemble several objects. There are a couple of ways you can begin a new assembly. If you are just starting SOLIDWORKS®, you can select the assembly icon in the *Welcome to SOLIDWORKS* window (see Figure 7.1-1). If you are already in SOLIDWORKS®, you can use the following methods to start a new assembly.

- Click on the **New** icon ▢ from the *Quick Access* toolbar at the top.
- Short cut key: **Ctrl + N**
- Pull-down menu: **File – New…**

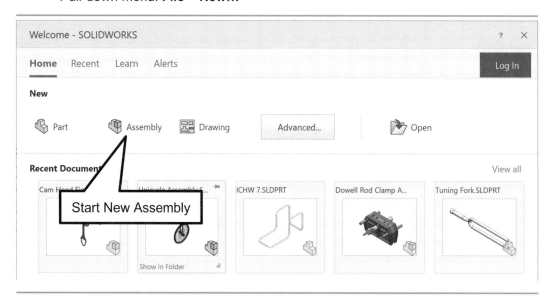

Figure 7.1-1: Welcome to SOLIDWORKS® - Home tab

7.1.2) Inserting Parts

There are three methods for inserting parts into an assembly. The method 1 is used if all or most of your parts included in the assembly are already opened. Method 2 is used if no parts are currently opened. Use method 3 for inserting additional parts.

<u>Inserting parts - Method 1 (Parts are open)</u>

1) **Open** all the parts that will be in the assembly.
2) Select a **File – New…**.
3) Select **Assembly**.
4) A *Begin Assembly* window will appear with all the open parts listed (See Figure 7.1-2). Highlight all the parts.
5) Move your cursor to the model area and click to insert the first part. Repeat until all the parts are inserted.

<u>Inserting parts - Method 2 (No parts are open)</u>

1) Select a **File – New…**.
2) Select **Assembly**.
3) An *Open* window will appear. Select all the parts that will be in the assembly.
4) A *Begin Assembly* window will appear with all the parts listed (See Figure 7.1-2).
5) Move your cursor to the model area and click to insert the first part. Repeat until all the parts are inserted.

<u>Inserting parts - Method 3 (Inserting additional parts)</u>

1) Click on the *Assembly* tab.

2) Select **Insert Components** .
3) An *Open* window will appear. Select the part(s) that are wanted to add to the assembly.
4) An *Insert Component* window will appear with the part(s) listed.
5) Move the cursor to the model area and click to insert the part. Repeat until all the parts are inserted.

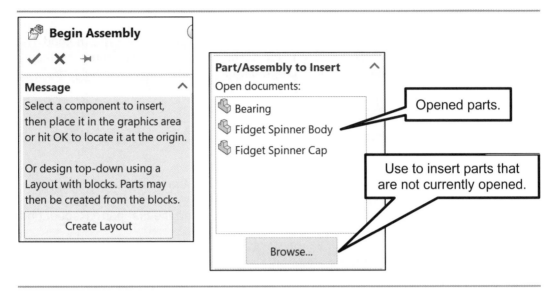

Figure 7.1-2: *Begin Assembly* Window

7.1.3) Feature Manager Design Tree

The *Assembly Feature Manager Design Tree*, shown in Figure 7.1-3, shows all the parts that make up the assembly. It also shows all the applied mates. These are the constraints that hold the assembly together. It also can be seen which of the parts are modeled parts and which parts came from the *Toolbox*.

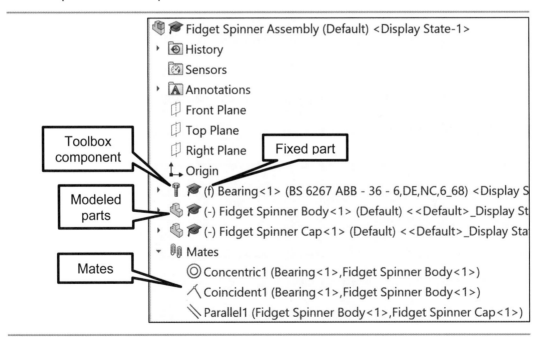

Figure 7.1-3: Feature Manager Design Tree

Another important thing to notice, in the *Feature Design Tree*, is the part that is fixed. SOLIDWORKS® will automatically fix one part. That means that the part is immovable. The fixed part will have an **(f)** in front of its name. To float (or un-fix a part) or fix a part use the following steps.

Floating or fixing a part

1) **Right-click** on the part you want to float/fix.
2) Select **Float** or **Fix**.

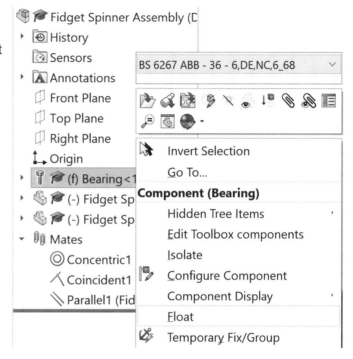

7.1.4) Copying Parts

If there are two or more identical parts or configurations of the same part, you can copy the part. Use the following steps to copy a part.

Copying a part

1) Hold the **Ctrl** key down.
2) **Click** on the part you wish to copy.
3) **Drag** the mouse to a new location and release.

If you make a copy of a part and there are different configurations of that part, use the following steps to select the appropriate configuration.

Selecting a configuration

1) **Click** on the part.
2) A pop-up will appear. From the pull-down menu, choose the correct configuration.

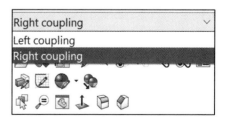

7.2) MATES

Mates allow physical constraints between parts in an assembly to be created. For example, making surface contacts or making two shafts run along the same axis. There are three categories of mates: *Standard, Advanced,* and *Mechanical*.

7.2.1) The Mate command

The **Mate** ⌗[Mate] command is located in the *Assembly* tab. Figure 7.2-1 shows the *Mate* options window and the available mates. The basic steps taken to apply a *Mate* are listed below.

Applying a Mate

1) Select the **Mate** ⌗[Mate] command. A *Mate* window will appear showing the available *Mates* (See Figure 7.2-1).
2) Select the two parts, features, surfaces, or planes to which you want the mate to be applied.
3) SOLIDWORKS will automatically select the mate that it thinks is most appropriate. If the default mate is good select **OK**, if not select the mate that you wish to apply.
4) If your part is oriented the wrong way, select one of the **Mate alignment** [Mate alignment icons] icons to flip the part.
5) The *Mate* window will stay active so that you can apply your next mate.
6) Select **OK** in the *Mate* window to end the *Mate* command.

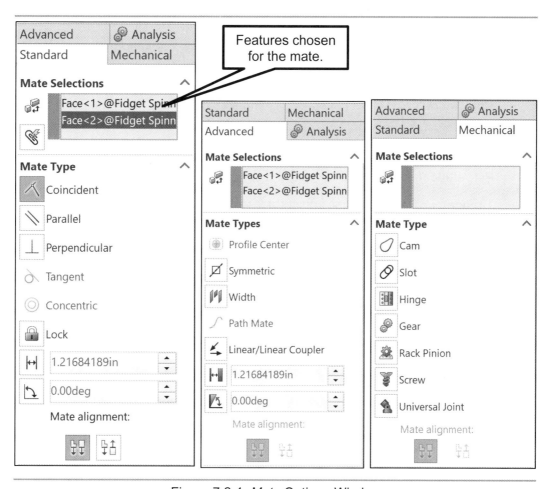

Features chosen for the mate.

Figure 7.2-1: *Mate* Options Window

7.2.2) Standard Mates

SOLIDWORKS® groups mates into three categories: *Standard* mates, *Advanced* mates, and *Mechanical* mates. This chapter will only deal with *Standard* mates. The following are descriptions of the available *Standard mates*.

- **Coincident:** Makes points, edges and surface occupy the same point, line or plane.

- **Parallel:** Makes edges and surfaces parallel.

- **Perpendicular:** Makes edges and surfaces perpendicular.

- **Tangent:** Makes edges and surfaces tangent.

- **Concentric:** Makes cylindrical surface and circular edges share the same center point.

- **Distance:** Make points, edges and surface a specified distance from each other.

- **Angle:** Make edges and surface a specified angle apart.

7.3) TOOLBOX

The **Toolbox** gives access to standard parts such as bolts, nuts, and gears. This saves the effort of having to model them. The **Toolbox** is located in

the *Design Library* . The *Design library* is located in the *Task Pane* on the right side of your drawing area.

7.3.1) Adding the Toolbox

The Toolbox is generally something that you have to Add-In. To preload the *Toolbox,* perform the following steps.

Adding in the Toolbox

1) Select **Tools – Add-ins….**
2) In the *Add-Ins* window, activate the **SOLIDWORKS Toolbox Library** and **SOLIDWORKS Toolbox Utilities** checkbox in the **Active Add-ins** column (See Figure 7.3-1). If you want the *Toolbox* to activate each time you open the program, activate it in the *Start Up* column as well.
3) Select **OK**.

7.3.2) Inserting a Toolbox Component

To access the *Toolbox* and insert a component, use the following generalized steps. If you use these steps and it does not allow you to insert a component, it is usually for one of two reasons. First, check to see if you are in an assembly file. You cannot insert a *Toolbox* component into a part file. Second, the *Toolbox* may not be added in yet. If which case, follow the steps listed above.

Inserting a Toolbox component

1) In the *Task Pane* on the right side of the screen, select **Design Library** 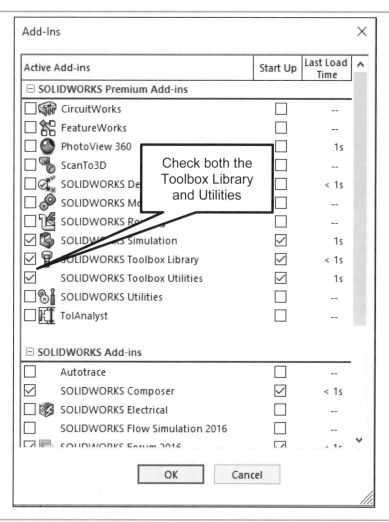.
2) **Expand** the *Toolbox* by clicking on the arrow to the left of the *Toolbox* name. If the Toolbox is not loaded, **Add in now**. (See Figure 7.3-2.)
3) **Expand** the standard that controls your model (e.g., ANSI inch, ANSI Metric).
4) **Expand** the category (e.g., Bearings, Bolts and Screws).
5) **Select** the type of part that you want to insert into your assembly (e.g., Hex Head, Lag Screws).
6) **Drag** the specific part to your assembly.
7) A window will appear allowing you to adjust the parameters of the part.
8) If you need to change the size of a toolbox component after it has been inserted, right-click on the component in the *Feature Manager Design Tree* and select **Edit Toolbox component**.

Figure 7.3-1: *Add-Ins* Window

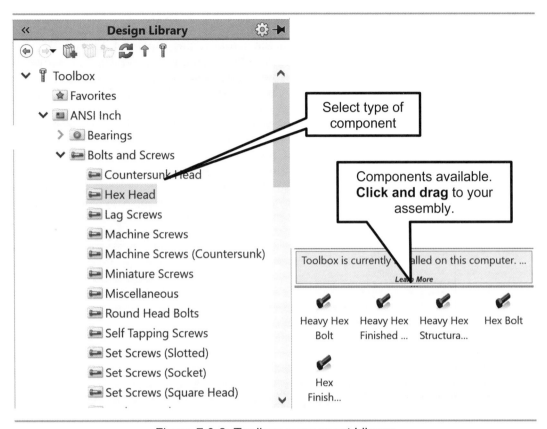

Figure 7.3-2: Toolbox component Library

7.4) FLANGED COUPLING ASSEMBLY TUTORIAL

7.4.1) Prerequisites

Before starting this tutorial, you should complete the following tutorials.

- Chapter 3 - Flanged Coupling Model Tutorial
- Chapter 5 – Flanged Coupling Configurations Tutorial

It will help if you have the following knowledge.

- Familiarity with assembly drawings
- Familiarity with threads and fasteners

7.4.2) What you will learn

The objective of this tutorial is to introduce you to simple assembly models. In this tutorial, you will be creating the remaining parts for the *Flanged Coupling* assembly shown in Figure 7.4-1. You will also assemble the parts using *Mates*. Specifically, you will be learning the following commands and concepts.

Setting up your drawing

- New assembly

Assembly

- Standard Mates
- Toolbox components
- Exploded view

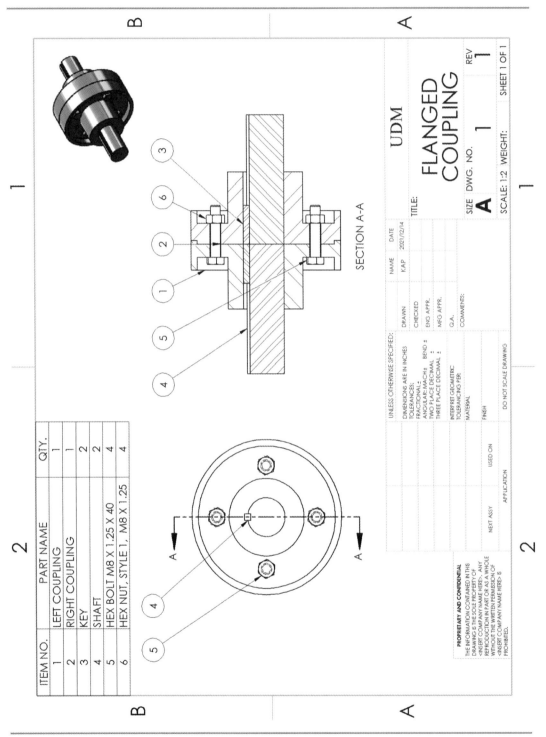

ITEM NO.	PART NAME	QTY.
1	LEFT COUPLING	1
2	RIGHT COUPLING	1
3	KEY	2
4	SHAFT	2
5	HEX BOLT M8 X 1.25 X 40	4
6	HEX NUT, STYLE 1, M8 X 1.25	4

SECTION A-A

	NAME	DATE
DRAWN	K.AP	2021/12/14
CHECKED		
ENG APPR.		
MFG APPR.		
Q.A.		
COMMENTS:		

UNLESS OTHERWISE SPECIFIED:

DIMENSIONS ARE IN INCHES
TOLERANCES:
FRACTIONAL±
ANGULAR: MACH±   BEND ±
TWO PLACE DECIMAL ±
THREE PLACE DECIMAL ±

INTERPRET GEOMETRIC
TOLERANCING PER:

MATERIAL

FINISH

DO NOT SCALE DRAWING

UDM

TITLE:

FLANGED
COUPLING

SIZE DWG. NO.   REV
A   1   1

SCALE: 1:2 WEIGHT: SHEET 1 OF 1

NEXT ASSY   USED ON

APPLICATION

Figure 7.4-1: Flanged Coupling

7.4.3) Shaft

1) Open a **new MMGS part** and save it as **SHAFT.SLDPRT** in the same folder as the *Coupling* created in Chapter 3.

2) **Sketch** and **Dimension** the following profile on the **Front plane**.

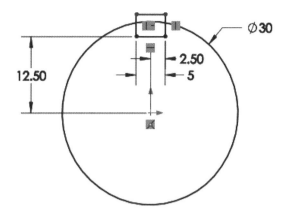

3) **Trim (Too closest)** 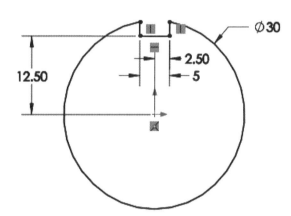 the excess lines to create the key seat.

> ➢ See Chapter 3 for information on ***Trimming***.

4) **Extrude** the sketch **50 mm** and make the part out of **AISI 1020** Steel.

5) **Save** the part.

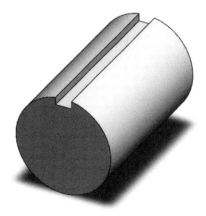

7.4.4) Key

1) Open a **new MMGS part** and save it as **KEY.SLDPRT** in the same folder as the *Coupling* created in Chapter 3.

2) **Sketch** and **Dimension** the following profile on the **Front plane**. Use a **Center Rectangle** [□ Center Rectangle].

3) **Extrude** the sketch **30 mm** and make the part out of **AISI 1020** Steel.

4) **Save** the part.

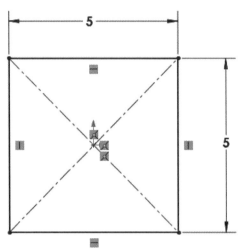

7.4.5) Setting up the assembly

1) Open **COUPLING.SLDPRT**, **SHAFT.SLDPRT**, and **KEY.SLDPRT**

2) Create a **new Assembly** .

> ➤ See section 7.1.1 for information on starting a *New Assembly*.

3) In the *Begin Assembly* window you should see all three of the parts in the *Open documents* field. Select all of the parts using either the *Shift* key or *Ctrl* key.

4) Move the mouse onto the drawing area. One of the parts will appear. Place the part by clicking the left mouse button. Another one of the parts will appear. Place this one and then next.

5) Note that all three parts appear in the *Feature Manager Design Tree*.

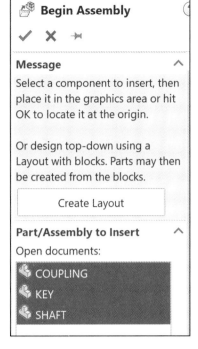

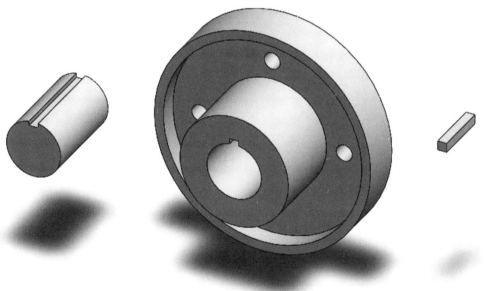

▸ 🦴 🎏 (f) KEY<1> (Default<<Default>_Display State 1>)

▸ 🦴 🎏 (-) SHAFT<1> (Default<<Default>_Display State 1>)

▸ 🦴 🎏 (-) COUPLING<1> (Left Coupling<<Default>_Display State 1>)

6) Make sure the units are **MMGS,** the standard is set to **ANSI**, and then save your assembly as **FLANGED COUPLING.SLDASM**.

7) Make a copy of each of the parts. Do this by holding the **Ctrl** key, clicking on the part and then dragging it to a new location.

8) Notice, in the *Feature Manager Design Tree*, it lists 6 parts (two couplings, two shafts and two keys). One of the parts has an (f) next to it. This is the fixed part (i.e., non-movable part). *Float* that part and *Fix* shaft<1>. Do this by right-clicking on the part and choosing either **Float** or **Fix**.

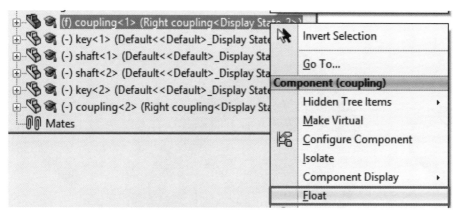

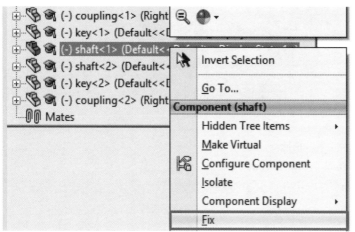

9) Notice that you either have two of the *left couplings* or two of the *right couplings*. We need one of each. To change the part's configuration, click on the part in the *Feature Manager Design Tree* and select the configuration. Visually confirm that you have one of each coupling configuration in the assembly. SOLIDWORKS® may have automatically chosen the non-represented configurations when you copied the coupling.

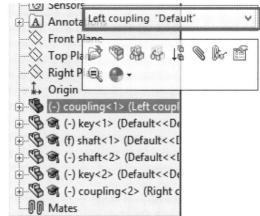

10) **Save All**. If a warning window appears, choose **Rebuild and save the document**.

7.4.6) Creating mates

1) **Hide** *shaft<2>*, *key<2>* and the *right coupling*. Do this by right-clicking on the part in the *Feature Manager Design Tree* and selecting **Hide Components** .

2) Apply a **Concentric Mate** [Mate] between the **Left Coupling's** center hole and **SHAFT<1>**'s circumference.

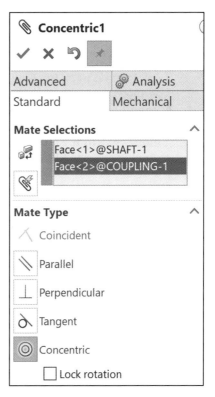

a) Select **Mate** [Mate] (*Assembly* tab).
b) Select the circumference of the *Shaft* and then the inside surface of the big hole in the *Coupling*.
c) Note that SOLIDWORKS® has automatically chosen the **Concentric** mate for you. Also note that the coupling has moved positions.
d) ✓
e) Notice that the *Mate* window is still open. You are ready to apply more mates.

> See section 7.2 to learn about *Mates*.

3) **Rotate** the view so that you can see the back of the *Coupling*. The *Coupling* may or may not be in the same location shown.

4) **Click and drag** the *Coupling*. Notice that it will only move on the axis of the *Shaft* because of the Mate that we just applied.

5) If there is any feature listed in the *Mate Selection* field, **right-click** on it and select **Clear Selections**.

6) Apply a **Coincident Mate** between the back edge of the **Left Coupling** and the back surface of **SHAFT<1>.** <u>Hint:</u> Select the two surfaces that you want to line up. You want the assembly to look like the figure. Remember to select the green check mark to apply the mate.

7) Notice how the cutout in the Shaft and the cutout in the Coupling may not line up. If it does line up, rotate the coupling to see that it is not fixed in this location. This next mate will fix that. Apply a **Coincident** mate between the **Right Plane** of the **Left Coupling**, and the **Right Plane** of the **SHAFT<1>**.

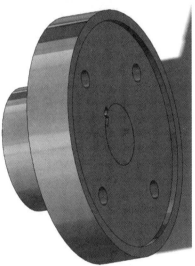

   a) Look to the right of the Mate window. You will see the *Feature Design Tree*. **Expand** it.
   b) **Expand** *Left Coupling*.
   c) **Expand** *Shaft 1*.
   d) **Select** the **Right Plane** under *Left Coupling*.
   e) **Select** the **Right Plane** under *Shaft 1*.
   f) ✓

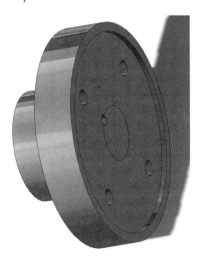

8) Create three **Coincident** mates that will place the **KEY** in its correct position. When done, the assembly should look like the figure.

9) Close ✅ the *Mate* window.

10) Look at the very bottom of the *Feature Design Tree*. You will see *Mates*. **Expand** the *Mates*. You should see the following mates or something very similar.

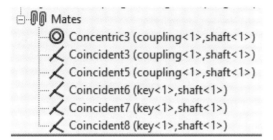

11) **Show Components** 👁 *SHAFT<2>, KEY<2>*

and the *right coupling*. **Hide Components** 🚫 *SHAFT<1>, KEY<1>* and the *left coupling*.

12) Make similar **Mates** 📎 between these three parts.

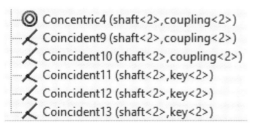

13) **Save** the assembly.

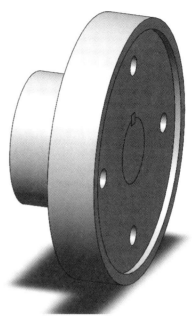

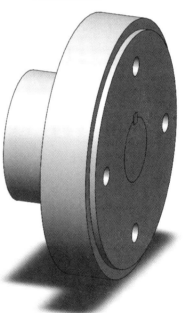

14) **Show Components** 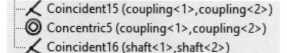 all of the parts and apply the following **Mates** .

a) **Coincident** mate between the contacting surfaces of the couplings. Flip the

Mate alignment if necessary.

b) **Concentric** mate between the outer diameters of each *Coupling*.

c) **Coincident** mate between the **Right Planes** of the two shafts. This will force the two couplings to align and rotate together.

> ✕ Coincident15 (coupling<1>,coupling<2>)
> ◎ Concentric5 (coupling<1>,coupling<2>)
> ✕ Coincident16 (shaft<1>,shaft<2>)

15) Notice that the shafts are not long enough. Something that we might not have known until the parts were assembled. In the *Feature Manager Design Tree*, **right-click** on one of the *Shafts* and select **Edit Part**.

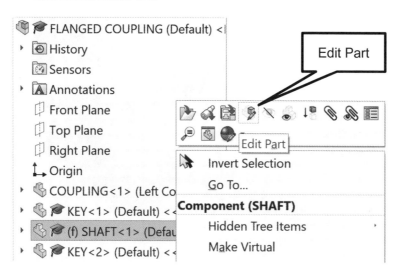

16) **Expand** the *Shaft*. Notice that its features are now written in blue.

17) Change the shaft length to **100 mm**. Do this by right-clicking on the **Extrude** feature and selecting **Edit Feature**.

18) **Right-click** on *Shaft* and select **Edit assembly**.

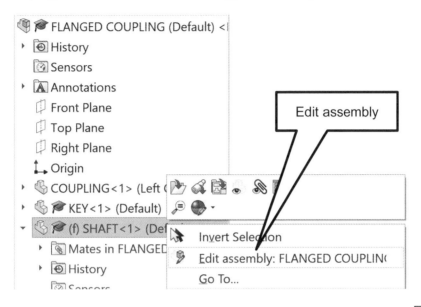

19) If the assembly file doesn't automatically update, select **rebuild** .

20) **Save all** the assembly.

7.4.7) The Toolbox

1) Click on the **SOLIDWORKS Add-Ins** tab. If the **SOLIDWORKS Toolbox** icon is not grey, click on it.

2) Insert **4X M8 Hex Bolt** with a length of **40 mm** into the assembly.

   a) In the *Task Pane* on the right side of the screen, select **Design Library** .

   b) **Expand** the *Toolbox* by clicking on the arrow to the left of the *Toolbox* name.

   c) Expand **ANSI Metric**.

   d) Expand **Bolts and Screws**.

   e) Select **Hex Head**.

   f) **Drag** the **Hex Bolt-ANSI B18** to the assembly.

   g) A *Configure Component* window will appear allowing you to adjust the parameters of the part. Input the following parameter.

       I.   Size = **M8**
       II.   Length = **40**
       III. Thread Length = **40**
       IV. Thread Display = **Simplified**

   h) ✅

   i) Click in the modeling area to place a total of **4** bolts.

   j) ❌

> ➤ See section 7.3 for information on the **Toolbox.**

3) Add **4X M8 Hex Nut Style 1** into the assembly with the following parameters.

   a) Size = **M8**

   b) Finish = **Double Chamfer**

   c) Thread Display = **Simplified**

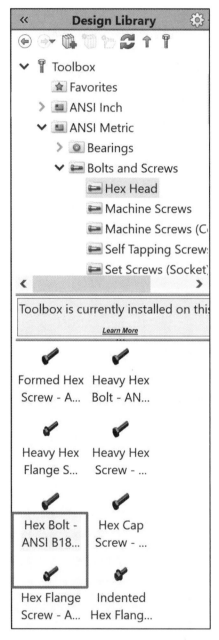

4) Apply the following **Mates** 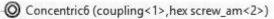 to put the bolts and nuts into position. Do this for every bolt-nut pair.

◎ Concentric6 (coupling<1>,hex screw_am<2>)
∠ Coincident17 (coupling<1>,hex screw_am<2>)
◎ Concentric7 (hex screw_am<2>,hex nut style 1_am<1>)
∠ Coincident18 (coupling<2>,hex nut style 1_am<1>)

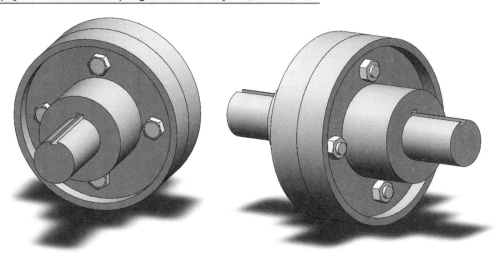

5) **Save all** the assembly.

7.4.8) Exploded assembly

1) **Explode** the assembly.   Create an explode assembly view similar to the one shown.

a) Select the **Exploded View** [Exploded View] command in the *Assembly* tab of the *Command Manager* ribbon.
b) Select a component and then click and drag on one of the axes to move it.
c) Select **Done** before you select the next component.
d) After you are done moving all of the parts into position, select ✔.
e) <u>Note:</u> You can also enter move values into the *Explode* window.

2) Enter the **Configurations** tab and expand the *Default* configuration. Notice that there is a new *Exploded View1* configuration. Double-click on **Exploded View1** to see the change.

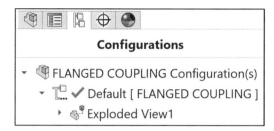

3) Make the unexploded view current. **Right-click** on *Exploded View1* and select **Animate explode**. Note that an *Animation Controller* will appear allowing you to manipulate the explode as well as to record it.

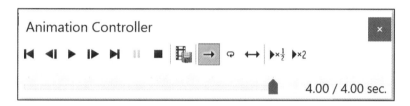

4) **Save**.

NOTES:

BASIC ASSEMBLIES IN SOLIDWORKS® PROBLEMS

P7-1) Use SOLIDWORKS® to create a solid model of the following parts. Create an assembly model of the *Drill Jig* applying the appropriate mates. E-mail the part files, including the toolbox components, and the assembly file to your instructor.

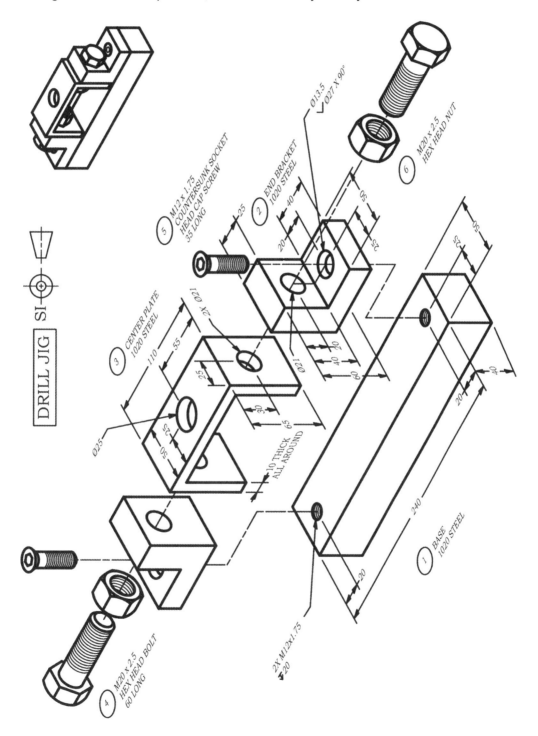

P7-2) Use SOLIDWORKS® to create a solid model of the following parts. Create an assembly model of the *Linear Bearing* applying the appropriate mates. E-mail the part files, including the toolbox components, and the assembly file to your instructor.

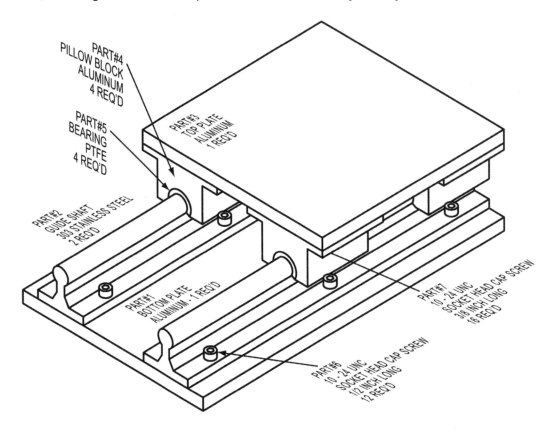

Part#1: Bottom Plate If you are just studying the basics and have not covered threads and fasteners, replace the 12X 10 – 24 UNC dimension with a 12x **n**.19 dimension.

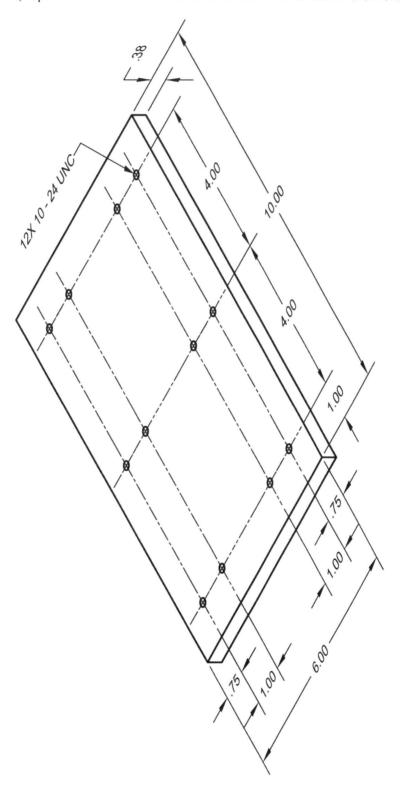

Part#2: Guide Shaft If you are just studying the basics and have not covered tolerancing, ignore the RC4 tolerance. NOTE TO DRAFTER: This part is symmetric, and all fillets are R.12.

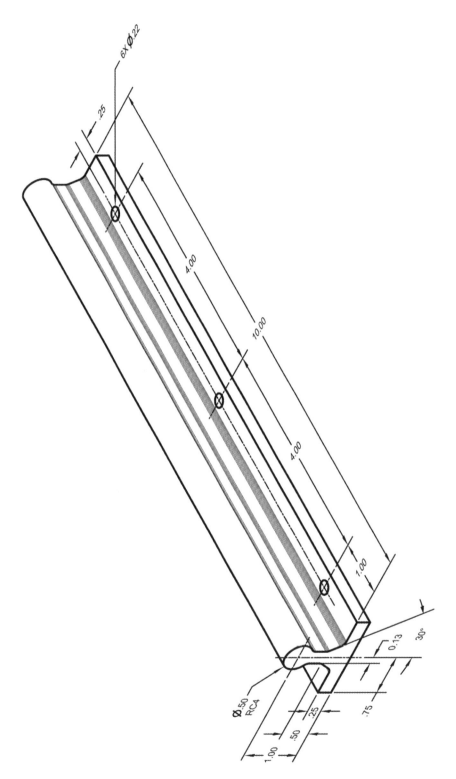

Part#3: Top Plate If you are just studying the basics and have not covered threads and fasteners, replace the 16X 10 − 24 UNC dimension with a 16x **n**.19 dimension.

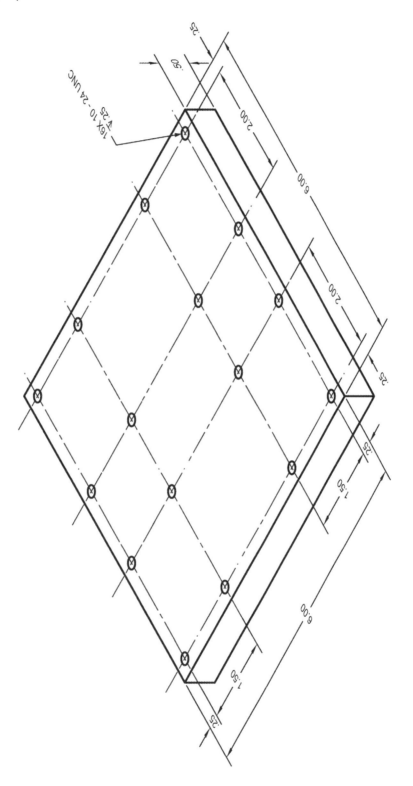

Part#4: Pillow Block If you are just studying the basics and have not covered tolerancing, ignore the FN1 tolerance.

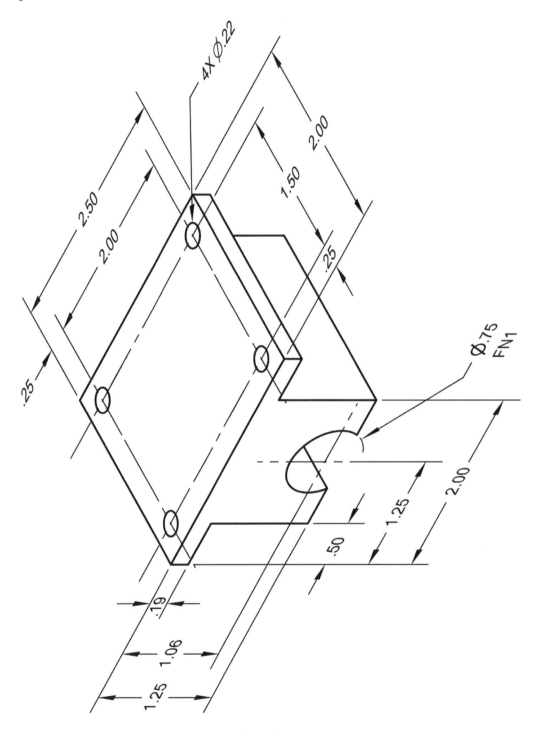

Part#5: Bearing If you are just studying the basics and have not covered tolerancing, ignore the RC4 and FN1 tolerances.

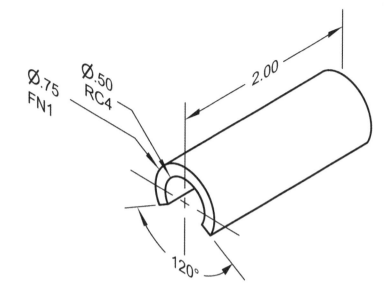

NOTES:

CHAPTER 8

ASSEMBLY DRAWINGS IN SOLIDWORKS®

CHAPTER OUTLINE

8.1) ASSEMBLY DRAWINGS 2

    8.1.1) What is an assembly drawing? 2

    8.1.2) How are parts in an assembly identified and located? 2

    8.1.3) How many parts are in an assembly? 2

8.2) BALLOONING 4

8.3) BILL OF MATERIAL 6

8.4) FLANGED COUPLING ASSEMBLY DRAWING TUTORIAL 8

    8.4.1) Prerequisites 8

    8.4.2) What you will learn 8

    8.4.3) Assembly views 10

    8.4.4) Bill of Materials 12

    8.4.5) Ballooning 14

ASSEMBLY DRAWINGS IN SOLIDWORKS® PROBLEMS 17

CHAPTER SUMMARY

In this chapter, you will learn how to construct an assembly drawing. You will also learn how to balloon an assembly and make a parts list. By the end of this chapter, you will be able to construct a complete and fully annotated assembly drawing.

8.1) ASSEMBLY DRAWINGS

8.1.1) What is an assembly drawing?

An assembly drawing is an orthographic projection of an entire machine or product. The purpose of an assembly drawing is to show how all the parts fit together and to suggest the function of the entire unit. Sometimes only one view is needed and sometimes it is necessary to draw all three principal views. It may also include specialized views such as section views or auxiliary views.

8.1.2) How are parts in an assembly identified and located?

A part is located and identified by using a circle or balloon containing a *find number* and a leader line that points to the corresponding part. A balloon containing a *find number* is placed adjacent to the part. A leader line, starting at the balloon, points to the part to which it refers. Balloons identifying different parts are placed in orderly horizontal or vertical rows. The leader lines are never allowed to cross and adjacent leader lines should be as parallel as possible to each other, as shown in Figure 8.1-1.

8.1.3) How many parts are in an assembly?

The parts list or bill of materials is an itemized list of all the parts that make up the assembled machine. Parts lists are arranged in order of their find number. Find and part numbers are usually assigned based on the size or importance of the part. The parts list is placed either in the upper left corner of the drawing, with part number 1 at the top, or lower right corner of the drawing, with part number 1 at the bottom (see Figure 8.1-1). A parts list may contain, but is not limited to, the following.

- Find number: The *find number* links the *parts list* description of the part to the balloon locating the part on the assembly drawing.
- Part number: A *part number* is an identifier of a particular part design. It is common practice, but not a requirement, to start the part number with the drawing number followed by a dash and then a unique number identifying the part. This unique number usually matches the find number. Standard parts usually do not contain the drawing number because they are used across designs, not just in a particular design.
- Nomenclature or description: The part name or description.
- Number of parts required (QTY REQ'D): The number of that part used in the assembly.
- Part material: The material the part is made of.
- Stock size: The pre-machined size of the part.
- Cage code: A cage code is a five-position code that identifies companies doing or wishing to do business with the Federal Government.
- Part weight: The weight of the finished part.

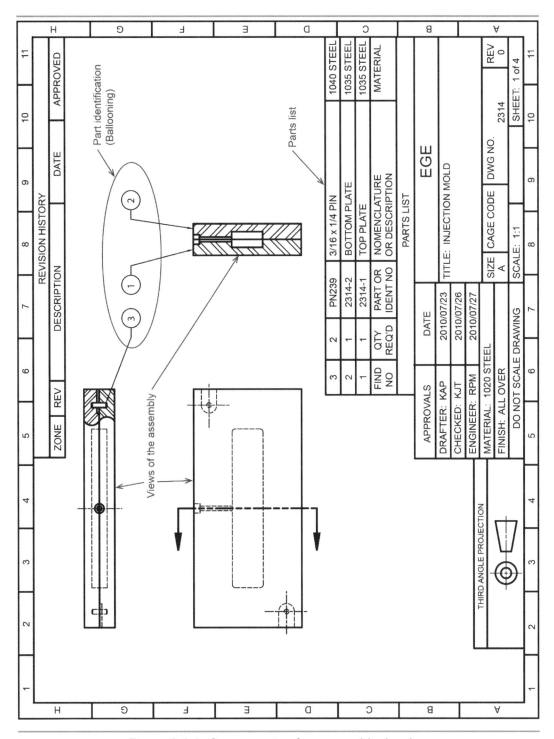

Figure 8.1-1: Components of an assembly drawing.

8.2) BALLOONING

Balloons are used to locate and identify each part in the assembly. The **Balloon** ⌖ Balloon command allows each balloon to be individually placed. The **Auto Balloon** ⌖ Auto Balloon command automatically chooses balloon location. The general steps for using the Balloon and Auto Balloon commands are given below. The option windows are shown in Figure 8.2-1.

Balloon command steps

1) Select the **Balloon** ⌖ Balloon command.
2) Characteristics of the balloons may be adjusted in the *Balloon* options window.
3) Select the part you wish to locate in the assembly.
4) Drag the balloon to the desired location.
5) If the find/part number within the balloon is not correct, you have not picked the correct part or the bill of materials is not correct.

Auto Balloon command steps

1) Select the **Auto Balloon** ⌖ Auto Balloon command.
2) Choose the balloon layout (e.g., horizontal, vertical).
3) Characteristics of the balloons may be adjusted in the *Auto Balloon* window.
4) Select the view where the balloons will be placed.

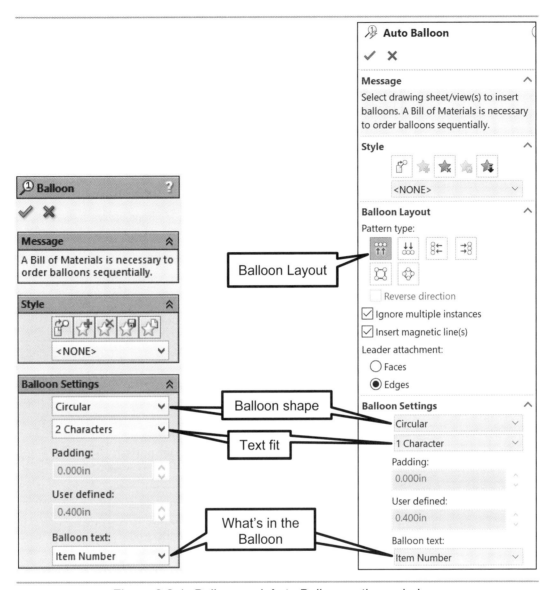

Figure 8.2-1: *Balloon* and *Auto Balloon* options windows

8.3) BILL OF MATERIAL

The **Bill of Material** 🔲 Bill of Materials command creates a table the contains a list of every part in the assembly. The *Bill of Materials* command is located in the *Annotation* tab stacked under *Tables*. Use the following steps to insert a bill of materials into the assembly drawing.

Tables	
⊞	General Table
🔲	Hole Table
🔲	Bill of Materials
🔲	Revision Table
🔲	Weld Table
🔲	Bend Table
🔲	Punch Table

Inserting a Bill of Material

1) Expand the **Table** icon in the *Annotation* tab and select **Bill of Material** 🔲 Bill of Materials.
2) Select a view that has all the parts represented.
3) Set the bill of material properties in the *Bill of Materials* options window (shown in Figure 8.3-2) and then select ✓.
4) Place the table. Usually in the upper left corner.
5) Editing text: The text may be edited by double clicking on it. Also, if a window shows up that asks about the link between the text and part, select **Keep Link**.
6) Sizing columns and rows: Click and drag on borders to resize columns and rows.
7) Reordering parts: Click and drag on a row to reorder the parts.

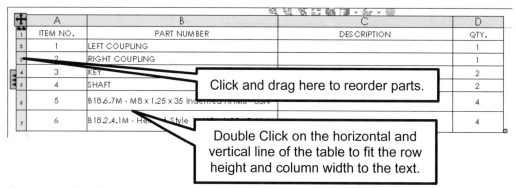

8) Edit table: Right-click on the table to access commands that allow for insertion or deletion of rows and columns, and other general formatting commands.
9) When the *Bill of Materials* table is opened, a window will appear that allows the table, cell and text properties to be adjusted.

10) When you double click on the text a window will appear that allows you to edit the text.

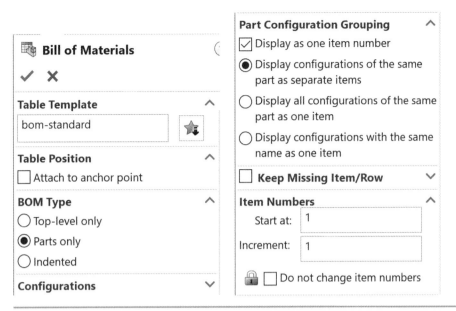

Figure 8.3-2: *Bill of Materials* options windows

8.4) FLANGED COUPLING ASSEMBLY DRAWING TUTORIAL

8.4.1) Prerequisites

Before starting this tutorial, you should have completed the following tutorials.

- Chapter 3 - Flanged Coupling Tutorial
- Chapter 7 - Flanged Coupling Assembly Tutorial

It will help if you have the following knowledge.

- Familiarity with assembly drawings

8.4.2) What you will learn

The objective of this tutorial is to introduce you to simple assembly drawings. In this tutorial, you will be creating the *Flanged Coupling* assembly drawing shown in Figure 8.4-1. Specifically, you will be learning the following commands and concepts.

Drawing

- Assembly drawing
- Ballooning
- Bill of Materials

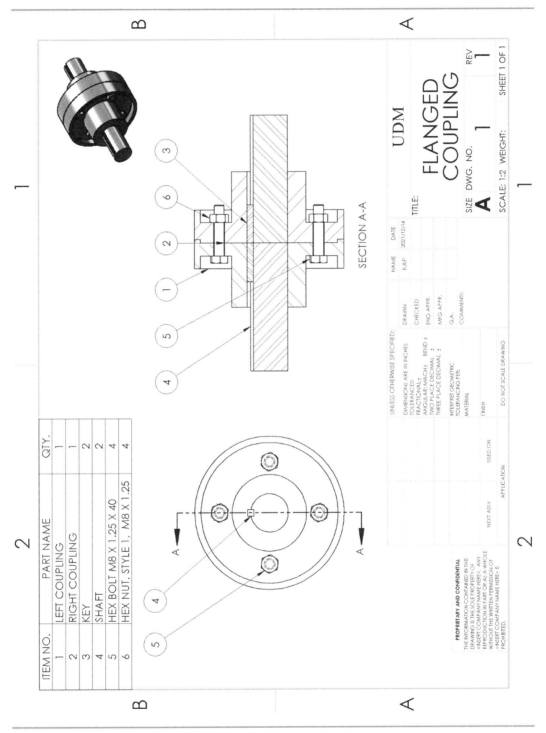

Figure 8.4-1: Flanged Coupling

8.4.3) Assembly views

1) **Open** the **FLANGED COUPLING.SLDASM** file that was created in Chapter 7.

2) Make sure that the assembly is in the non-exploded state.

3) Start a **New Drawing** .
   a) Select an **A (ANSI) Landscape** paper. (If you only see ISO paper sizes, deselect *Only show standard formats.*)
   b) Double click on **FLANGED COUPLING** in the *Model View* window.
   c) Make sure the *Show in exploded or model break state* check box is <u>deselected.</u> Create a front view that is at a **1:2** scale and a pictorial that is at a **1:4** scale.
   d) Set the drawing standard to **ANSI,** use **All uppercase for notes, tables, dimensions, and hole callouts**.

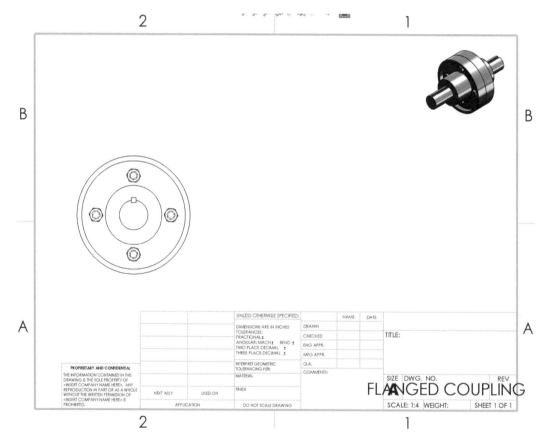

4) **Save** your drawing as **FLANGED COUPLING**.

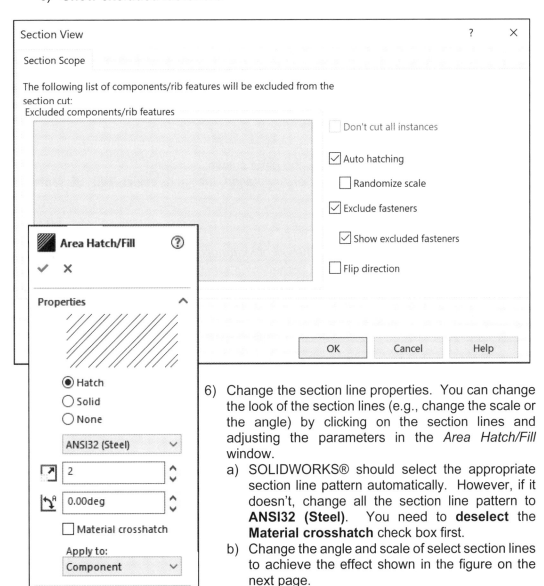

5) Create a **Section View** (located in the *View Layout* tab) that looks like the section view shown on the next page. Note that the section lines, in the section view, run in different directions and have different scales. This is to show physically different parts. When you create the section view, activate the following options.
   a) **Auto hatching**
   b) **Exclude fasteners** (We don't want to section the fasteners.)
   c) **Show excluded fasteners**

6) Change the section line properties. You can change the look of the section lines (e.g., change the scale or the angle) by clicking on the section lines and adjusting the parameters in the *Area Hatch/Fill* window.
   a) SOLIDWORKS® should select the appropriate section line pattern automatically. However, if it doesn't, change all the section line pattern to **ANSI32 (Steel)**. You need to **deselect** the **Material crosshatch** check box first.
   b) Change the angle and scale of select section lines to achieve the effect shown in the figure on the next page.

7) Add **Centerlines** where necessary.

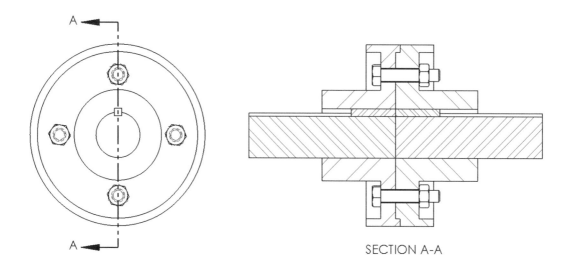

SECTION A-A

8.4.4) Bill of Materials

1) Add a **Bill of Material** that contains the parts shown.
   a. Expand the **Table** icon in the *Annotation* tab and select **Bill of Material**
      Bill of Materials
   b. Select the section view.

   c. Except the default selection in the *Bill of Materials* options window and select ✓
   d. Place the table in the upper left corner.
   e. Your table may look something like what is shown. There are a few problems with it that we need to fix.

> See section 8.3 to learn about the **Bill of Materials** command.

ITEM NO.	PART NUMBER	DESCRIPTION	QTY.
1	COUPLING		1
2	KEY		2
3	SHAFT		2
4	COUPLING		1
5	B18.2.3.5M - HEX BOLT M8 X 1.25 X 40 --40N		4
6	B18.2.4.1M - HEX NUT, STYLE 1, M8 X 1.25 --D-N		4

2) Reorder the parts. When you hover over the table a blue border appears. Grab on a blue border number and either drag up or down to reorder the parts. We want the more important parts to be first. Your table should look like the one shown when finished.

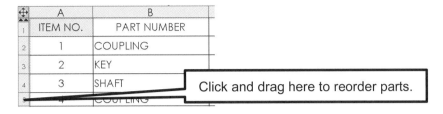

ITEM NO.	PART NUMBER	DESCRIPTION	QTY.
1	COUPLING		1
2	COUPLING		1
3	SHAFT		2
4	KEY		2
5	B18.2.3.5M - HEX BOLT M8 X 1.25 X 40 --40N		4
6	B18.2.4.1M - HEX NUT, STYLE 1, M8 X 1.25 --D-N		4

3) Note that both Couplings are named the same thing. We need to change that, but we don't know which one is which. So, we will do that later. Note: your couplings may already be named differently.

4) Change the fastener text to something simpler. Double-click on the text to edit. Select **Yes** when asked. Also, adjust the column width to fit the text.

5	HEX BOLT M8 X 1.25 X 40
6	HEX NUT, STYLE 1, M8 X 1.25

5) Right-click somewhere in the *DESCRIPTION* column and select **Select – Column**. Then, hit the **Delete** key.

6) Change the PART NUMBER text to **PART NAME**.

7) Double-click on row and column lines to auto-fit to the contents.

ITEM NO.	PART NAME	QTY.
1	COUPLING	1
2	COUPLING	1
3	KEY	2
4	SHAFT	2
5	HEX BOLT M8 X 1.25 X 40	4
6	HEX NUT, STYLE 1, M8 X 1.25	4

8.4.5) Ballooning

1) Use the **Auto Balloon** command to add the part identification.

    a) Select the **Auto Balloon** [Auto Balloon] command, located in the Annotation tab.

    b) Select the **Horizontal on Top** [icon] *Balloon Layout*.

    c) Use a **Circular** balloon with **2 Characters**.

    d) You can grab on a balloon and move the entire row if it needs to be repositioned.

    e) [✓]

    f) You can move the arrows if they don't point to the part in a clear way.

> ➢ See section 8.2 to learn about the *Balloon* command.

2) If a Balloon ended up in your pictorial, delete it.

3) Use the **Balloon** command to add any missing part identifications.

    a) Select the **Balloon** [Balloon]

    b) Use a **Circular** balloon with **2 Characters**.

    c) Select a part and then pull out the Balloon and place it in line with the row of other Balloons.

4) Continue to delete and add balloons to achieve the results shown in the figure on the next page.

5) Now that we know that Part 1 is the *Left Coupling* and Part 2 is the *Right Coupling*, we can change the text in the *Bill of Materials*.

6) Fill in the Title Block with the appropriate information.

7) **Save**.

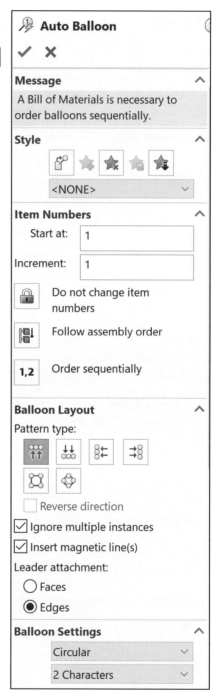

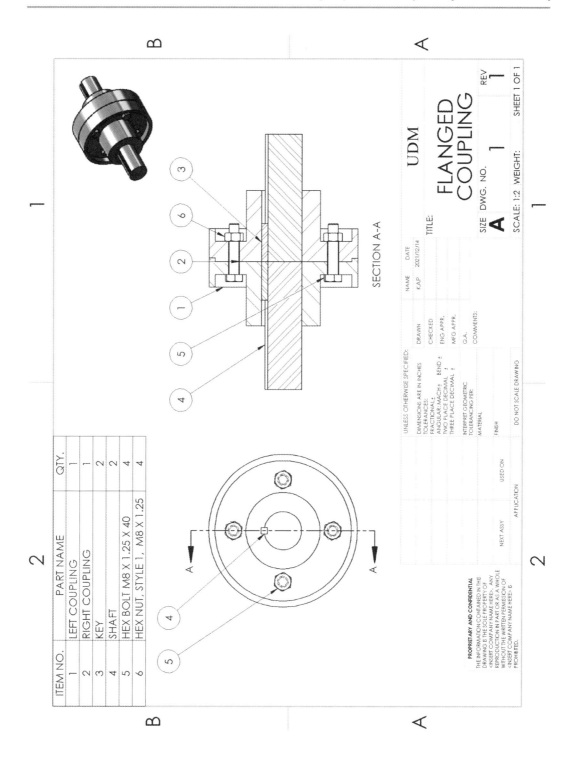

ITEM NO.	PART NAME	QTY.
1	LEFT COUPLING	1
2	RIGHT COUPLING	1
3	KEY	2
4	SHAFT	2
5	HEX BOLT M8 X 1.25 X 40	4
6	HEX NUT, STYLE 1, M8 X 1.25	4

SECTION A-A

UNLESS OTHERWISE SPECIFIED:

DIMENSIONS ARE IN INCHES
TOLERANCES:
FRACTIONAL±
ANGULAR: MACH±   BEND ±
TWO PLACE DECIMAL   ±
THREE PLACE DECIMAL ±

INTERPRET GEOMETRIC
TOLERANCING PER:

MATERIAL

FINISH

DO NOT SCALE DRAWING

	NAME	DATE
DRAWN	K.AP	2021/12/14
CHECKED		
ENG APPR.		
MFG APPR.		
Q.A.		
COMMENTS:		

UDM

TITLE:

FLANGED COUPLING

SIZE **A**   DWG. NO.   1   REV 1

SCALE: 1:2   WEIGHT: SHEET 1 OF 1

NEXT ASSY   USED ON

APPLICATION

NOTES:

ASSEMBLY DRAWINGS IN SOLIDWORKS® PROBLEMS

P8-1) Use SOLIDWORKS® to reproduce the assembly drawing of the *Drill Jig* shown. The part dimensions are given in Chapter 7.

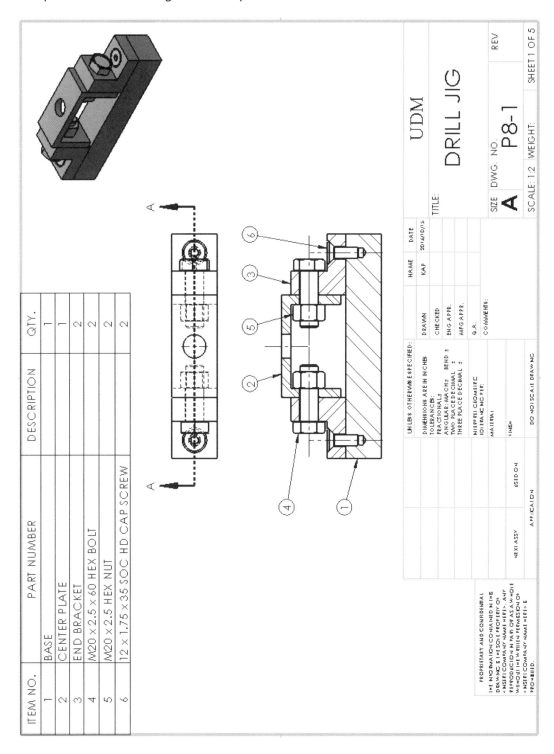

ITEM NO.	PART NUMBER	DESCRIPTION	QTY.
1	BASE		1
2	CENTER PLATE		1
3	END BRACKET		2
4	M20 x 2.5 x 60 HEX BOLT		2
5	M20 x 2.5 HEX NUT		2
6	12 x 1.75 x 35 SOC HD CAP SCREW		2

P8-2) Use SOLIDWORKS® to reproduce the assembly drawing of the *Linear Bearing* shown. The part dimensions are given in Chapter 7.

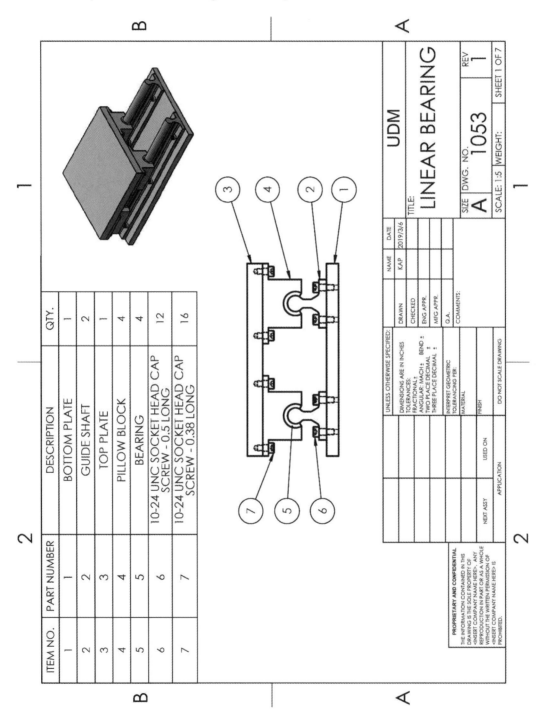

ITEM NO.	PART NUMBER	DESCRIPTION	QTY.
1	1	BOTTOM PLATE	1
2	2	GUIDE SHAFT	2
3	3	TOP PLATE	1
4	4	PILLOW BLOCK	4
5	5	BEARING	4
6	6	10-24 UNC SOCKET HEAD CAP SCREW - 0.5 LONG	12
7	7	10-24 UNC SOCKET HEAD CAP SCREW - 0.38 LONG	16

UNLESS OTHERWISE SPECIFIED:

DIMENSIONS ARE IN INCHES
TOLERANCES:
FRACTIONAL±
ANGULAR: MACH± BEND ±
TWO PLACE DECIMAL ±
THREE PLACE DECIMAL ±

INTERPRET GEOMETRIC
TOLERANCING PER:

MATERIAL

FINISH

DO NOT SCALE DRAWING

	NAME	DATE
DRAWN	KAP	2019/3/6
CHECKED		
ENG APPR.		
MFG APPR.		
Q.A.		
COMMENTS:		

TITLE:

UDM

LINEAR BEARING

SIZE **A** | DWG. NO. **1053** | REV **1**

SCALE: 1:5 | WEIGHT: | SHEET 1 OF 7

NEXT ASSY | USED ON

APPLICATION

CHAPTER 9

ADVANCED PART MODELING IN SOLIDWORKS®

CHAPTER OUTLINE

9.1) SKETCHING 3

    9.1.1) Sketch 3

    9.1.2) Arcs 4

    9.1.3) Conics 4

    9.1.4) Converting Entities, Intersection Curve, and Silhouette Entities 9

    9.1.5) Mirror Entities 10

    9.1.6) Scale 11

9.2) FEATURES 12

    9.2.1) Loft 12

    9.2.2) Dome 14

    9.2.3) Shell 16

    9.2.4) Rib 17

    9.2.5) Sweep 18

    9.2.6) Reference Geometry 21

9.3) MICROPHONE BASE TUTORIAL 22

    9.3.1) Prerequisites 22

    9.3.2) What you will learn 22

    9.3.3) Setting up the arm 24

    9.3.4) Loft 24

    9.3.5) Base joint 28

    9.3.6) Shell 32

    9.3.7) Ribs 33

9.4) MICROPHONE ARM TUTORIAL 36

    9.4.1) Prerequisites 36

    9.4.2) What you will learn 36

    9.4.3) Setting up the arm 36

    9.4.4) Loft 36

    9.4.5) Shell the arm 41

    9.4.6) The arm joint 42

9.5) BOAT TUTORIAL 45

9.5.1) Prerequisites 45

9.5.2) What you will learn... 45

9.5.3) Setting up the project... 46

9.5.4) Loft 46

9.5.5) Shell, sweep and rib... 50

ADVANCED PART MODELING IN SOLIDWORKS® PROBLEMS..**55**

CHAPTER SUMMARY

In this chapter you will learn how to create complex shapes in SOLIDWORKS®. Sketch elements such as conics, construction elements, Convert Element, and Intersection Curve, will be explored. Ways of modifying sketches will also be used such as Mirror Entity and Scale. Features such as Loft, Shell, Rib, Sweep, Dome, and Mirror will be explored. By the end of this chapter, you will be able to create complex curved shapes in SOLIDWORKS®.

9.1) SKETCHING

9.1.1) Sketch

In Chapter 1 and 3 we discussed several different sketch entities. In this chapter, we will practice making **Arcs** and **Ellipses**. We will also be modifying sketch entities using the **Mirror Entities** and **Scale Entities** commands. The *Scale Entities* is buried under the *Move Entities* command. Many different sketch elements are available in the ***Sketch* tab** shown in Figure 9.1-1.

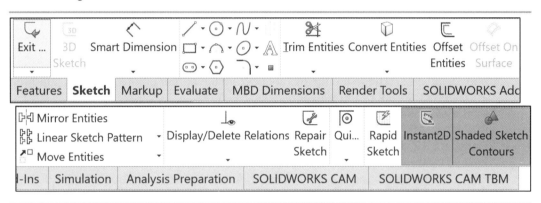

Figure 9.1-1: Sketch tab commands

9.1.2) Arcs

You are able to draw different arc types. The arc commands are located in the **Sketch tab** in the *Command Manager*. A description of each arc type is given below.

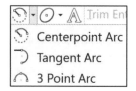

- **Centerpoint Arc**: Defined by a center point, a starting point and an endpoint.
- **3 Point Arc**: Defined by a starting point, endpoint and a radius.
- **Tangent Arc**: Defined by the endpoint of another entity that it will be tangent to and an endpoint.

9.1.3) Conics

A conic is a curve obtained by intersecting a cone with a plane. Conics available in SOLIDWORKS® include the shapes: *Ellipse*, *Partial Ellipse*, *Parabola* and *Conic*. These commands are described below.   For steps on how to create these sketches, see the informational boxes.

- **Ellipse:** This command creates an ellipse. An ellipse is a regular oval shape. It is defined by a center point and the locations of the major and minor axes. These are the distances from the center point to the long quadrant and short quadrant of the ellipse, respectively.
- **Partial Ellipse:** This command allows an ellipse to be created with part of it missing.
- **Parabola:** This command creates a parabola. A parabola is a symmetric curve formed by the intersection of a cone with a plane.
- **Conic:** This command creates a conic. A conic is like a parabola, but it doesn't have to be symmetric. A conic gives flexibility when creating the curve.

To draw an *Ellipse,* use the following steps. The *Ellipse* dimension may be adjusted in the *Ellipse* option window shown in Figure 9.1-2. The *Ellipse* also can be dimensioned using *Smart Dimensions*.

Drawing an *Ellipse*

1) Select **Ellipse** ⊙ Ellipse .
2) Select a center point.
3) Select the location of either the major or minor axis.
4) Select the location of the other axis.

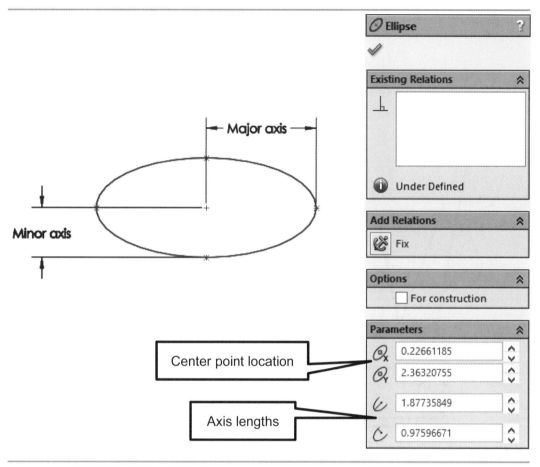

Figure 9.1-2: Ellipse option window

To draw a *Partial Ellipse,* use the following steps. The *Partial Ellipse* dimension may be adjusted in the *Ellipse* option window shown in Figure 9.1-3. The *Partial Ellipse* can also be dimensioned using *Smart Dimensions.*

Drawing a *Partial Ellipse*

1) Select **Partial Ellipse** $\widehat{C}$ **Partial Ellipse** .
2) Select a center point.
3) Select the location of either the major or minor axis.
4) Select the location of the other axis.
5) Select the extents of the *Partial Ellipse.*

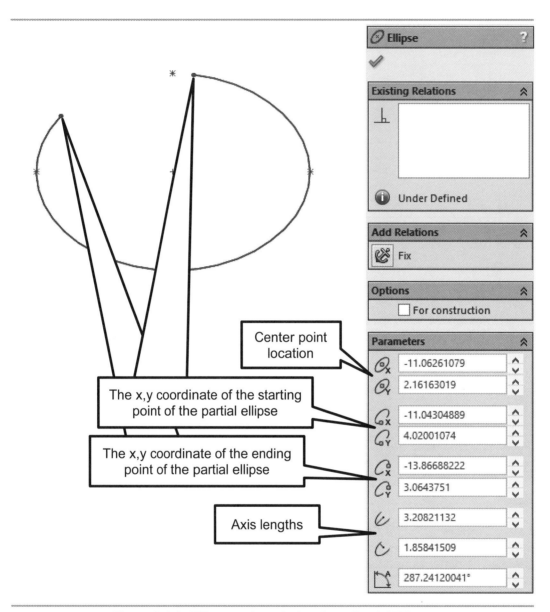

Figure 9.1-3: Ellipse option window for a Partial Ellipse

To draw a *Parabola,* use the following steps. The *Parabola* dimension may be adjusted in the *Parabola* option window shown in Figure 9.1-4. The *Parabola* can also be dimensioned using *Smart Dimensions*.

Drawing a *Parabola*

1) Select **Parabola** ⎣∪ Parabola⎦.
2) Select a focal point.
3) Select the vertex.
4) Select the extents of the *Parabola*.

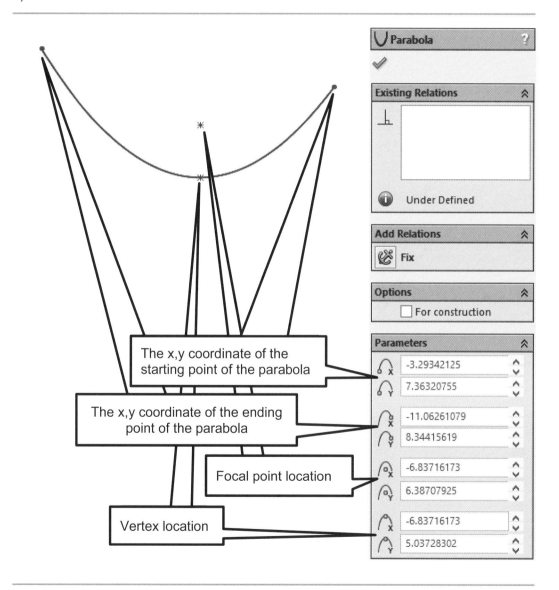

Figure 9.1-4: Parabola option window

To draw a *Conic,* use the following steps. The *Conic* dimension may be adjusted in the *Conic* option window shown in Figure 9.1-5. The *Conic* can also be dimensioned using *Smart Dimensions*.

Drawing a *Conic*

1) Select **Conic** ⟨∩ Conic⟩.
2) Select the start point.
3) Select the end point.
4) Select the top vertex.
5) Select the shoulder point.

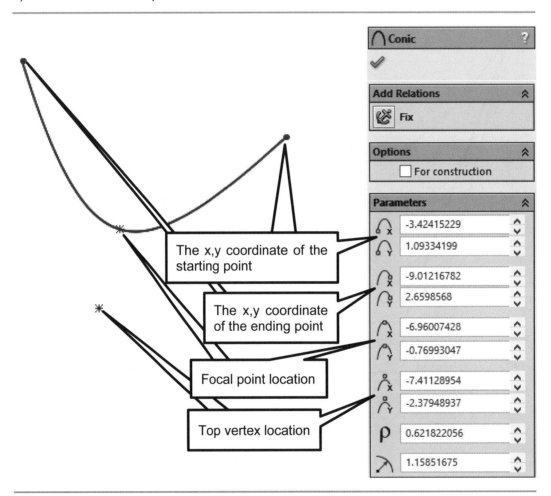

Figure 9.1-5: Conic option window

9.1.4) Converting Entities, Intersection Curve, and Silhouette Entities

The *Convert Entities* command allows you to create one or more curves by projecting an edge, face, or curve of an existing solid or sketch onto a plane. The *Intersection Curve* command allows you to create a sketch based off the intersection of a solid geometry and a plane. Other intersections, such as between two surfaces, a surface and a model face may be used. The Silhouette Entities command allows you to create sketches by projecting the outline of a part onto a sketch plane. Examples of uses of each command is given in Figure 9.1-6. For the steps on how to use these commands, see below.

Creating a sketch using *Converting Entities*

1) Select the **Convert Entities** 　Convert Entities command.
2) Select the face, edge or curve that needs to be converted.
3) Select OK.

Creating a sketch using *Intersection Curve*

1) Sketch on the plane or face where the sketch will be supported.
2) Select the **Intersection Curve** 　Intersection Curve command.
3) Select the faces or edges that will be used to create the intersecting sketch.
4) Select OK.

Creating a sketch using *Silhouette Entities*

1) Sketch on the plane or face where the sketch will be supported.
2) Select the **Silhouette Entities** 　Silhouette Entities command.
3) Select the solid that will be used to create the sketch.
4) Select OK.

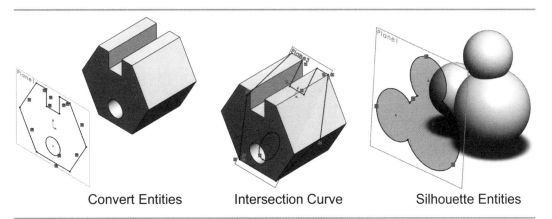

Convert Entities   Intersection Curve   Silhouette Entities

Figure 9.1-6: Examples of projecting curves

9.1.5) Mirror Entities

The *Mirror Entities* command mirrors 2D sketches about a chosen line. To use the *Mirror Entities* command, use the following steps. Figure 9.1-7 shows the *Mirror* options window.

Mirroring sketches

1) Select the **Mirror Entities** ⊞ Mirror Entities command.
2) Select the entities that need to be mirrored.
3) Select the mirror line.
4) Choose whether you want to copy the entities or replace them.

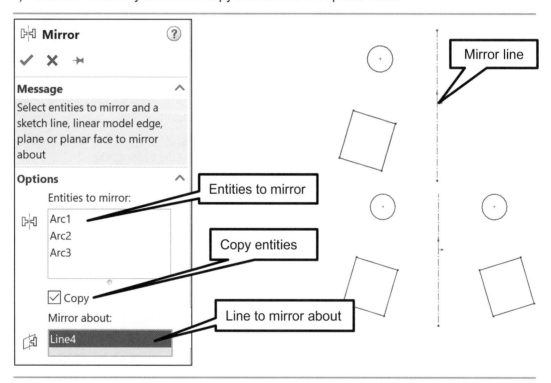

Figure 9.1-7: Mirror option window

9.1.6) Scale

The *Scale Entities* command allows you to uniformly change the size of sketch entities. To use the *Scale Entities* command, use the following steps. Figure 9.1-8 shows the *Scale* options window.

Scaling a sketch

1) Select the **Scale Entities** ⌐↗ Scale Entities command.
2) Select the entities to scale.
3) Select the point to scale about.
4) Enter the scale. The amount you want to increase or decrease the scale of the entities.
5) Select if you want to copy the entities. If the copy toggle is checked, the original entities will not be deleted.

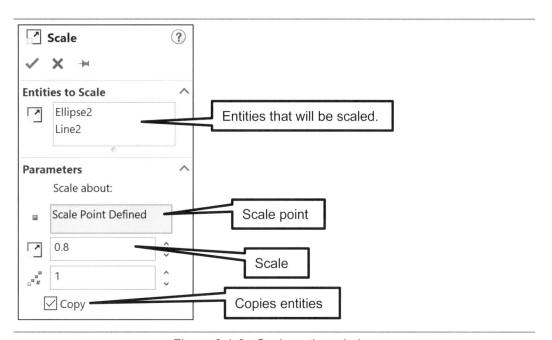

Figure 9.1-8: Scale option window

9.2) FEATURES

In Chapter 1 and 3, the feature-based commands of *Extrude, Extrude Cut, Fillet, Chamfer, Revolve, Hole, Mirror,* and *Pattern* were introduced. In this chapter, we will introduce the **Loft, Shell, Rib, Sweep,** *and* **Dome** commands. The use of **Reference Geometries** will also be expanded. The feature commands are located in the ***Feature tab*** shown in Figure 9.2-1.

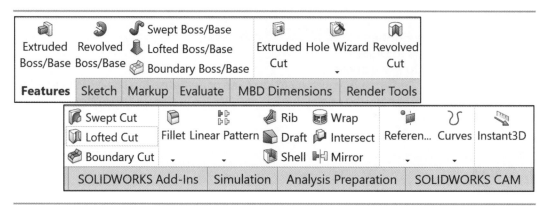

Figure 9.2-1: Feature tab commands

9.2.1) Loft

Lofts create or eliminate solids between two or more sketches on different planes. The solid/cut takes on the shape of the sketches as it proceeds from one to the other. A guide curve may also be used to give direction to the loft. Figure 9.2-2 shows examples of a solid created using the *Loft* command.

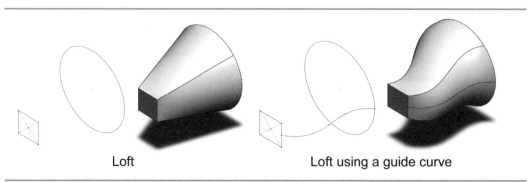

Loft   Loft using a guide curve

Figure 9.2-2: Examples of the *Loft* command

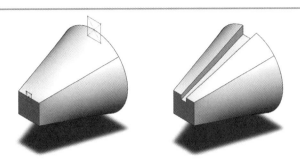

Figure 9.2-3: Example of the *Loft Cut* command

Use the following steps to create a Lofted Boss/Base or Lofted Cut. The Loft option window is shown in Figure 9.2-4.

1) Select the **Lofted Boss/Base** ⬇ Lofted Boss/Base or **Lofted Cut** 🔲 Lofted Cut command.
2) Select the sketches you wish to include in the loft. Note: If the loft is twisted, it is because you selected the sketches in the wrong order or because of the location that you selected in each sketch. You can adjust the twist by moving the green circles that appear.
3) Select the guide curve if one is being used.

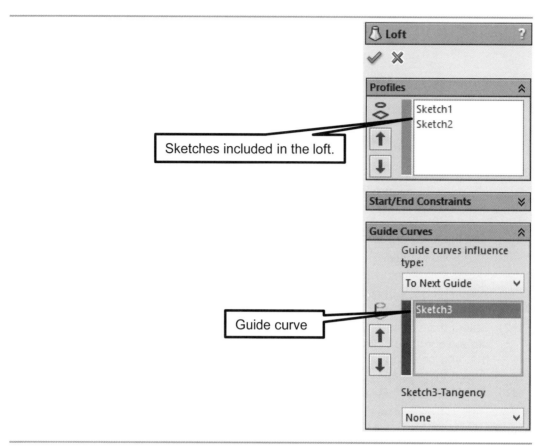

Figure 9.2-4: Loft option window

9.2.2) Dome

A *Dome* adds a rounded solid onto a face of an existing solid. The Dome command, by default, is not part of the Feature tab. However, it can be accessed through the pull-down-menu (**Insert-Feature-Dome**). Figures 9.2-5 through 9.2-7 show examples of different dome configurations. Figure 9.2-8 shows the *Dome* option window.

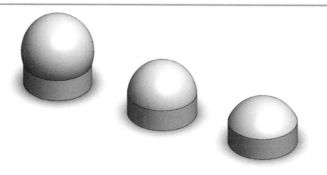

Figure 9.2-5: Domes with different radii

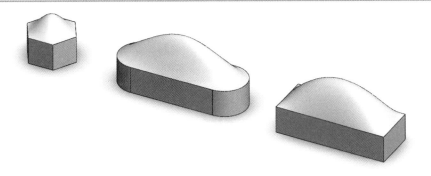

Figure 9.2-6: Domes on different shapes

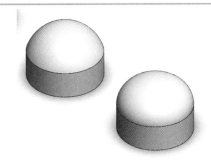

Figure 9.2-7: Example of a *Continuous Dome* (left) and an *Elliptical Dome* (right) of equal radius

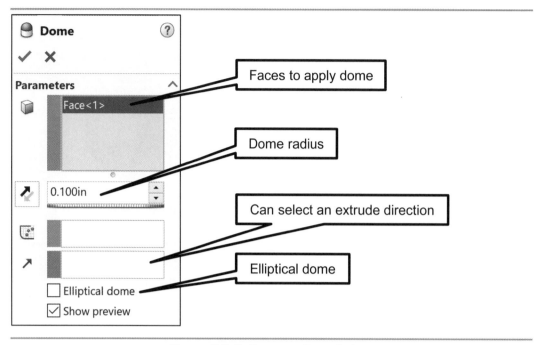

Figure 9.2-8: *Dome* option window

9.2.3) Shell

The *Shell* [Shell] command allows a solid part to be hollowed out. One or more faces may be removed in the process. Figure 9.2-9 shows examples of the *Shell* command. Figure 9.2-10 shows the *Shell* option window.

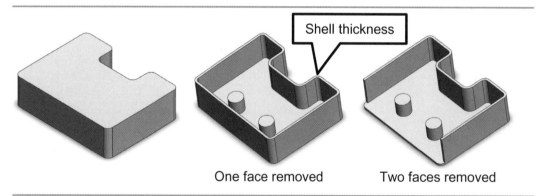

One face removed   Two faces removed

Figure 9.2-9: Examples of the *Shell* command

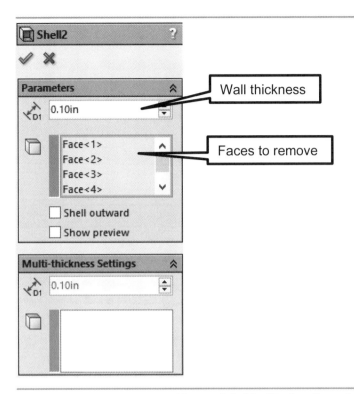

Figure 9.2-10: *Shell* option window

9.2.4) Rib

A *Rib* is a support structure. It is used to strengthen a part that is otherwise brittle or flimsy. It is usually thin relative to its length. A rib can be created using a single sketch entity. Figure 9.2-11 shows the example of the *Rib* command.

Figure 9.2-11: Examples of the *Rib* command

Use the following steps to create a *Rib*. Figure 9.2-12 shows the *Rib* option window.

Creating a Rib

1) Select the **Rib** command [🪶 Rib].
2) Select the sketch that will be used to create the rib(s).
3) Select the rib thickness method. The options are to add material on one side only or both sides.
4) Enter the rib thickness.
5) Select the direction of rib travel.
6) A draft may be applied to the rib.

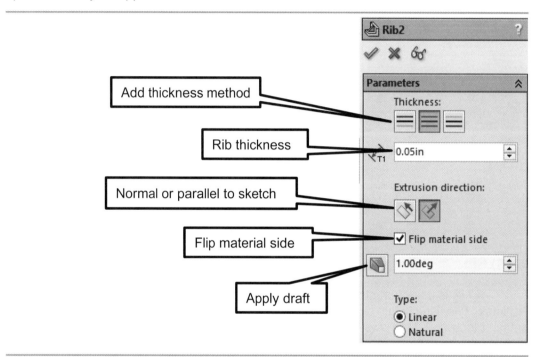

Figure 9.2-12: Rib option window

9.2.5) Sweep

The *Sweep* commands create or remove material by sweeping a profile along a path. Several options are available that allow you to dictate how the profile follows the path. Figure 9.2-13 and 9.2-14 show examples of the *Swept* and *Swept Cut* commands, respectively.

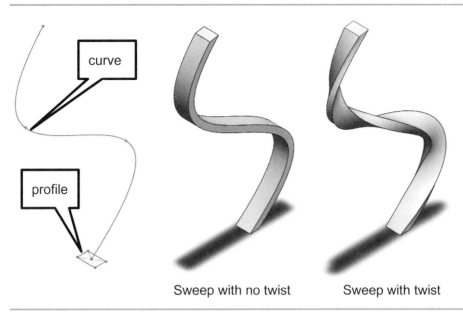

Sweep with no twist   Sweep with twist

Figure 9.2-13: Examples of the *Swept* command

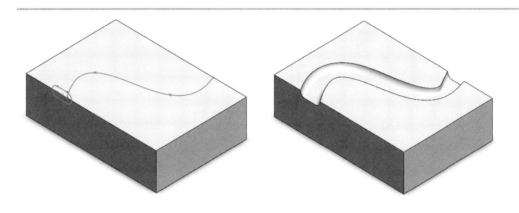

Figure 9.2-14: Examples of the *Swept Cut* command

Use the following steps to create a sweep. The Sweep option window is shown in Figure 9.2-15.

Creating a Sweep

1) Select the **Swept Boss/Bass** 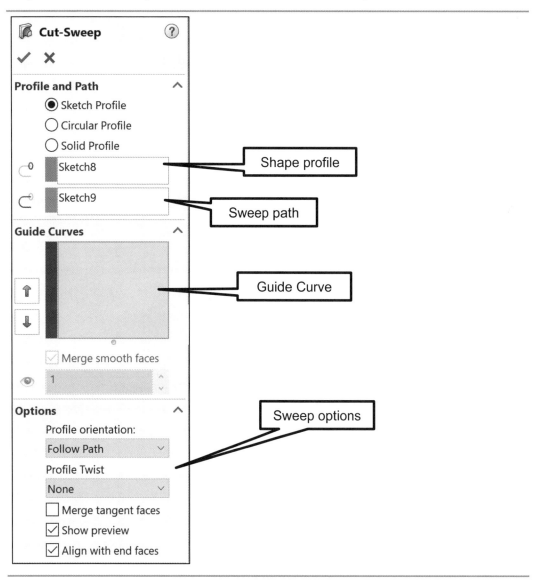 Swept Boss/Base or **Swept Cut** Swept Cut command.
2) Select the profile. This is the shape of the sweep.
3) Select the sweep path.
4) Select any sweep options. To get an idea of what each option affects, it is best to practice with each option and see how it effects the sweep.

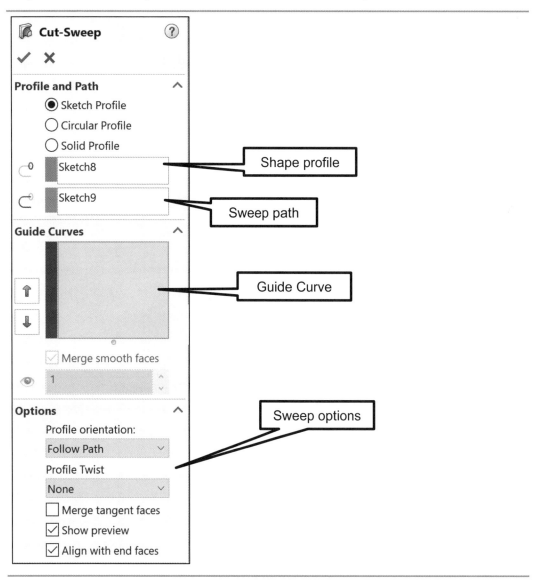

Figure 9.2-15: Sweep option window

9.2.6) Reference Geometry

Reference Geometries are used as drawing aids. They are not directly used to create solid geometry. The reference geometries that are available in SOLIDWORKS® are *Plane*, *Axis*, *Coordinate System*, *Point*, *Center of Mass*, and *Mate Reference*. There are many ways to define location and orientation of the reference geometries. There are too many to cover all of them here. The best way to learn the different methods is to practice with the options. If there is a particular orientation that is becoming difficult to create, go to the SOLIDWORKS® help forum, the videos that come with this book, or YouTube®.

- **Plane:** Creates additional planes parallel or at an angle to an existing plane or face.
- **Axis:**
- **Coordinate System:**
- **Point:**
- **Center of Mass:**
- **Mate Reference:**

9.3) MICROPHONE BASE TUTORIAL

9.3.1) Prerequisites

Before starting this tutorial, you should have completed the following tutorials.

- Chapter 3 – Tabletop Mirror Tutorial

9.3.2) What you will learn

The objective of this tutorial is to introduce you to creating more complex solid geometries. You will be modeling the *Base* of the *Microphone* assembly shown in Figure 9.3-1. Specifically, you will be learning the following commands and concepts.

Sketch

- Arc
- Convert Entities
- Mirror entities
- Construction Geometry

Feature

- Loft
- Mirror
- Shell
- Rib
- Reference Geometry (e.g., Plane, Axis, Coordinates)

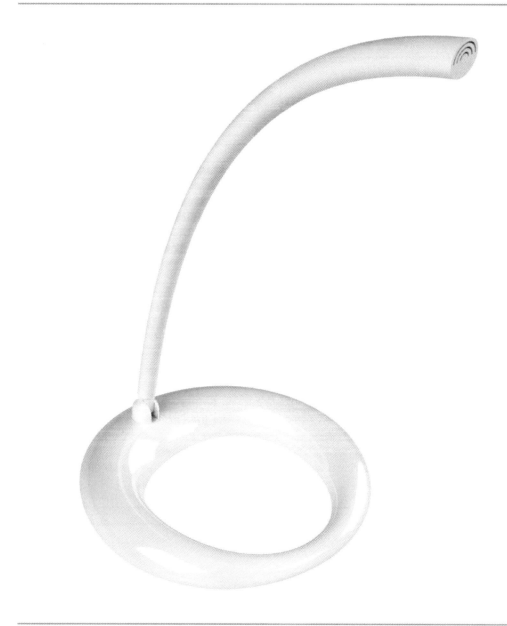

Figure 9.3-1: Microphone Assembly

9.3.3) Setting up the arm

1) Start a **new part** .

2) Set the drafting standard to **ANSI** and set all the text to **Upper case for notes, tables, and dimensions**. (*Options* 🔘 *– Document Properties – Drafting Standard*)

3) Set the units to **IPS** (i.e. inch, pound, second) and set the **Decimals = .123**. Also, select the rounding option, **Round half to even**. (*Options* 🔘 *– Document Properties – Units*)

4) Save the part as **MICROPHONE BASE.SLDPRT** (**File – Save**). Remember to save often throughout this project.

9.3.4) Loft

1) **Sketch** `Sketch` and **Dimension** `Smart Dimension` the following three profiles (use **Line** `Line` and **3 Point Arc** `3 Point Arc`). The first sketch is on the **Right Plane**, the second is on the **Front Plane**, and the third is on the **Right Plane**. Note that each sketch is separate (**Exit Sketch** between each) and that the line and origin are aligned horizontally. If the arcs are not black, the *sketch relations* may need to be manually applied. Remember that **Ctrl + 8** will give the normal view of the sketch plane and **Ctrl + 7** gives the isometric view.

> ➤ See section 9.1.2 to learn about *Arcs*.

First Sketch (Right Plane)

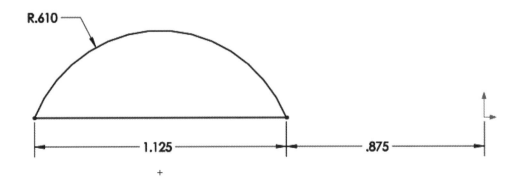

Second Sketch (Front Plane)

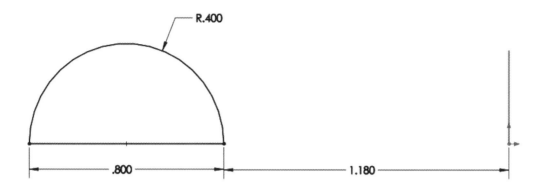

R.400

.800   1.180

Third Sketch (Right Plane)

R.188

1.625   .375

Final result (Hit Ctrl+7 to view the isometric)

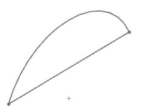

2) **Sketch** $\boxed{\text{Sketch}}$ and **Dimension** $\boxed{\text{Smart Dimension}}$ the following semi-circle (use **3 Point Arc** $\boxed{\text{3 Point Arc}}$) on the **Top Plane**. Note that the ends of the arc are **Coincident** with the center of the two existing *Right Plane* sketches. If your arc is not black, you may have to manually add the *Sketch Relations*.

3) **Exit** the **Sketch**.

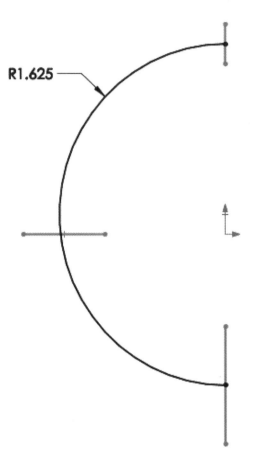

R1.625

4) **Deselect** all sketches and then select the **Loft** Lofted Boss/Base command. Select the three line/arc sketches using the semi-circle as the guide curve. Activate the **Normal To Profile** constraint for both the start and end.

> ➤ See section 9.2.1 to learn about *Lofts*.

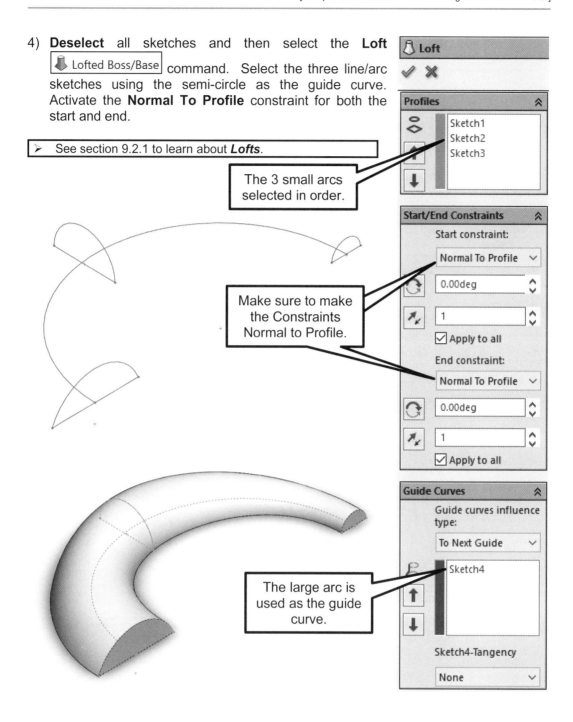

The 3 small arcs selected in order.

Make sure to make the Constraints Normal to Profile.

The large arc is used as the guide curve.

Loft

Profiles
Sketch1
Sketch2
Sketch3

Start/End Constraints
Start constraint:
Normal To Profile
0.00deg
1
☑ Apply to all
End constraint:
Normal To Profile
0.00deg
1
☑ Apply to all

Guide Curves
Guide curves influence type:
To Next Guide
Sketch4
Sketch4-Tangency
None

5) **Mirror** 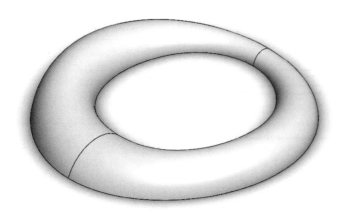 the *Loft* about the **Right Plane**.

9.3.5) Base joint

1) Create a **Plane** 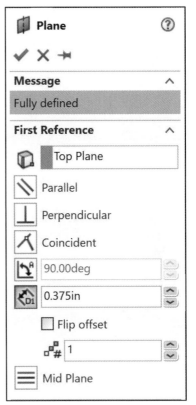 that is offset by **0.375 in** from the **Top Plane**.

> ➤ See section 9.2.6 to learn about **Reference Geometry**.

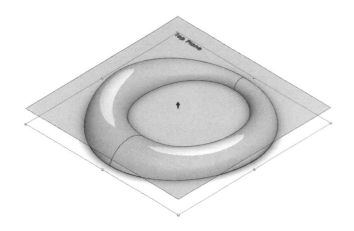

2) **Sketch** 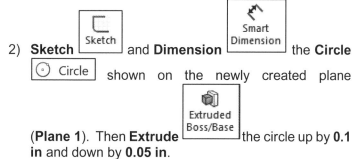 and **Dimension** the **Circle** shown on the newly created plane

(**Plane 1**). Then **Extrude** the circle up by **0.1 in** and down by **0.05 in**.

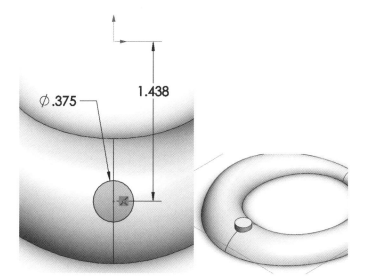

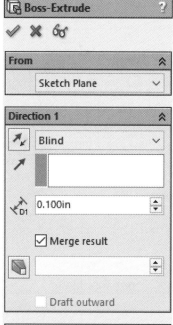

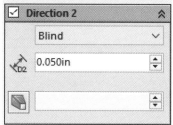

3) Add a **Dome** (**Insert – Feature – Dome**) to the top of the *Extrude* with a radius of **0.1875 in**. Select the flat surface of the extrusion to place the *Dome*.

> ➤ See section 9.2.2 to learn about *Domes*.

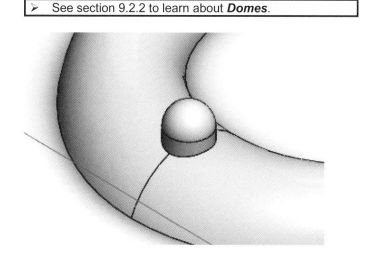

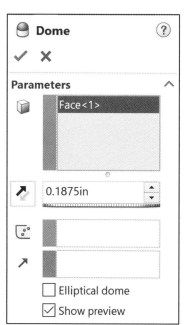

4) **Sketch** 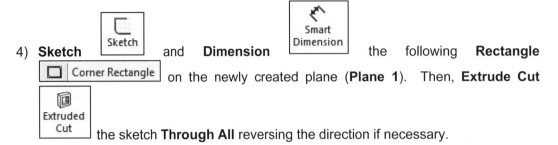 and **Dimension** the following **Rectangle** on the newly created plane (**Plane 1**). Then, **Extrude Cut** the sketch **Through All** reversing the direction if necessary.

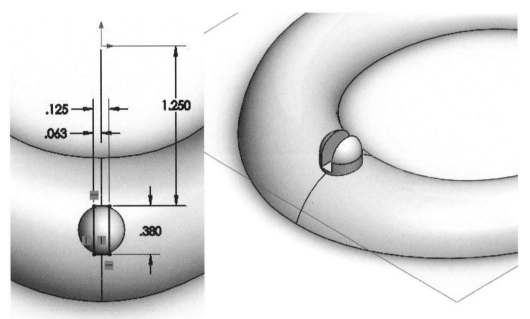

5) **Sketch** 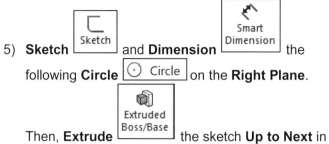 and **Dimension** the following **Circle** on the **Right Plane**.

Then, **Extrude** the sketch **Up to Next** in both directions. Note that the center of the circle is coincident with the center of the dome.

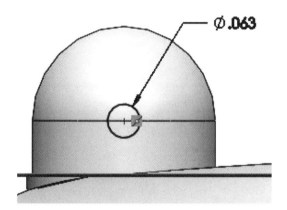

Ø .063

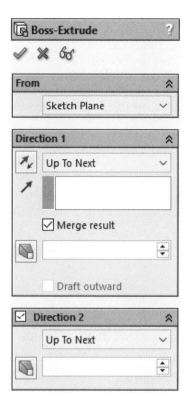

Boss-Extrude

From
Sketch Plane

Direction 1
Up To Next

☑ Merge result

☐ Draft outward

☑ **Direction 2**
Up To Next

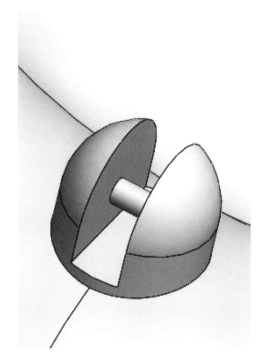

9.3.6) Shell

1) We would like to *Shell* the base, but not the base joint. In order to accomplish this, we will **Roll Back** the *Feature Manager Design Tree* to a place the *Shell* before we started to model the joint. Notice the line at the bottom of the *Design Tree*. Hover the mouse over this line. When a hand appears, click and drag the line until it is under **Mirror1**.

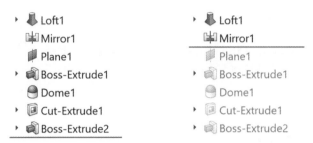

2) **Shell** the base to a thickness of **0.1 in** removing the bottom surface by selecting the bottom surface when you apply the *Shell*.

> ➢ See section 9.2.3 to learn about ***Shells***.

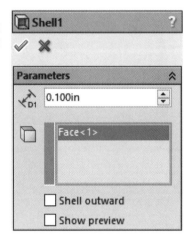

3) **Roll Forward** the *Design Tree* until after the last step.

4) We will not be needing **Plane1** anymore so **Hide** 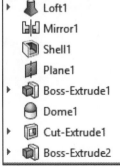 it.

9.3.7) Ribs

1) **Sketch** [Sketch] on the **Top Plane**. View the part from the bottom. **Convert Entities** [Convert Entities] the edge shown in the figure and change the elements indicated in the figure into a **Construction Geometry** [icon].

> ➤ See section 9.1.4 to learn about *Converting Entities*.

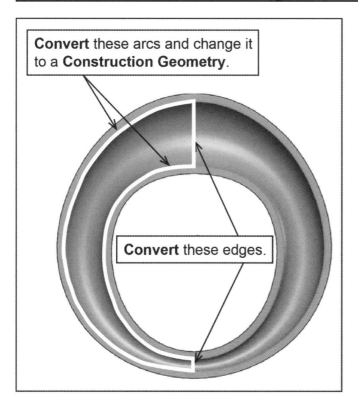

Convert these arcs and change it to a **Construction Geometry**.

Convert these edges.

2) Draw and **Dimension** [Smart Dimension] the three **Centerlines** [Centerline] and **Arcs** [3 Point Arc] shown in the figure. Note that the endpoints of each arc are **Coincident** with the construction geometry arcs. Make sure all the arcs are black, which means that they are completely constrained. **Mirror** [Mirror Entities] the three arcs using one of the converted lines to mirror about. Select the entities first before selecting the *Mirror* command.

> ➤ See section 9.1.5 to learn about the ***Mirror Entities*** command.

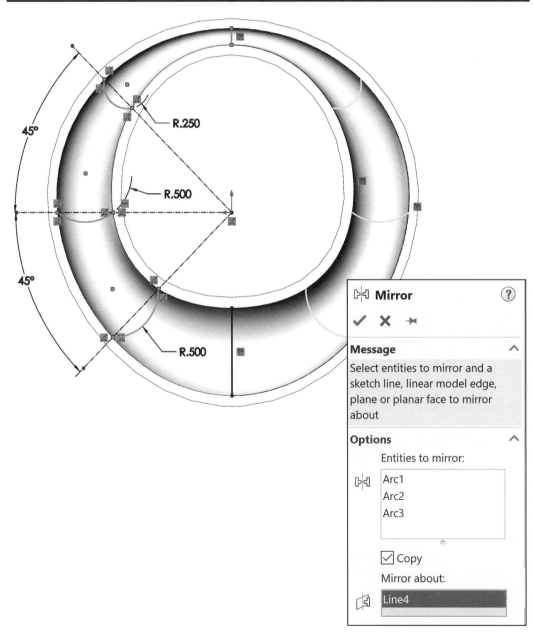

3) **Rib**  the sketch using a thickness of **0.05 in**. Add thickness to both sides. Flip the material side if needed.

> ➢ See section 9.2.4 to learn about *Ribs*.

4) Make the Base out of **PVC Rigid**.

5) **IMPORTANT!!! Save** this part and **keep** it for use in a future tutorial.

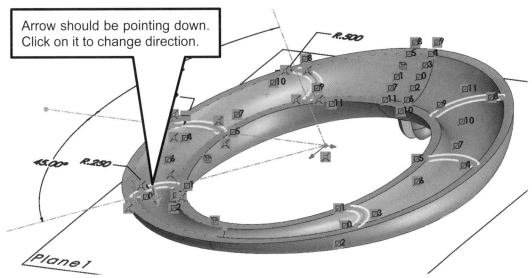

9.4) MICROPHONE ARM TUTORIAL

9.4.1) Prerequisites

Before starting this tutorial, you should have completed the following tutorial.

- Microphone Base Tutorial

9.4.2) What you will learn

The objective of this tutorial is to introduce you to creating more complex solid geometries. You will be modeling the *Arm* of the *Microphone* assembly shown in Figure 9.3-1.

9.4.3) Setting up the arm

1) Start a **new part** Part .

2) Set the drafting standard to **ANSI** and set the text to **Upper case for notes, tables, and dimensions**. (*Options* ⚙ – *Document Properties – Drafting Standard*)

3) Set the units to **IPS** (i.e. inch, pound, second) and set the **Decimals = .123**. Also, select the rounding option, **Round half to even**. (*Options* ⚙ – *Document Properties – Units*)

4) Save the part as **MICROPHONE ARM.SLDPRT** (**File – Save**). Remember to save often throughout this project.

9.4.4) Loft

1) **Sketch** and **Dimension** the following **Circle** on **Top Plane**. **Exit Sketch**.

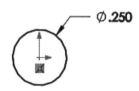

Ø.250

2) Create a **Plane**  that is **4.5 in** from the **Right Plane**.

3) **Sketch** and **Dimension** the following **Circle** on **Plane1** (the newly created plane). Make the circle center and origin aligned **Vertically**. **Exit Sketch**.

4) **Sketch** and **Dimension** the following **Centerline** on the **Front Plane**. Note that the end of the *Centerline* is aligned **Horizontally** with the origin and **Coincident** with Plane1. **Exit Sketch**.

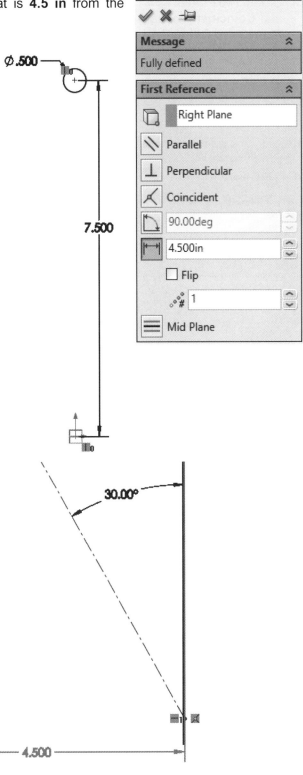

5) Create two **Planes** 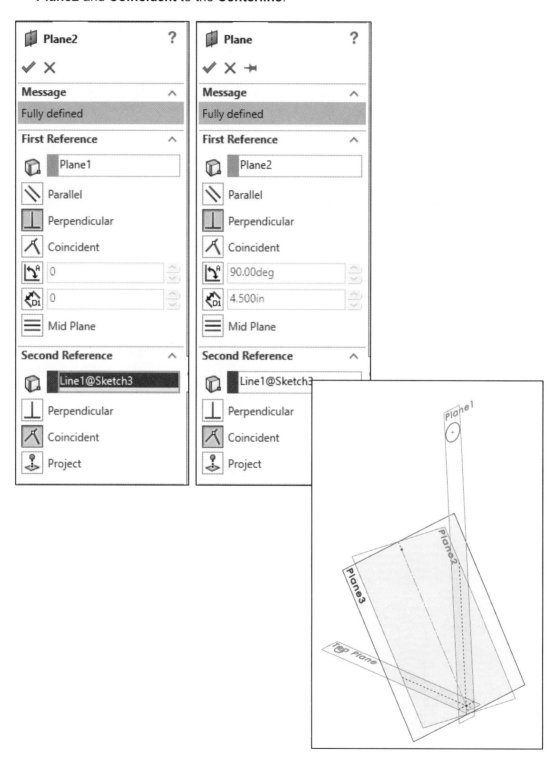. For the first plane, make it **Perpendicular** to **Plane1** and **Coincident** to the **Centerline**. For the second plane, make it **Perpendicular** to **Plane2** and **Coincident** to the **Centerline**.

6) **Sketch** and **Dimension** the following **Circle** on **Plane3** (the last plane created). **Exit Sketch**. Note that the center of the circle is **Coincident** with the centerline.

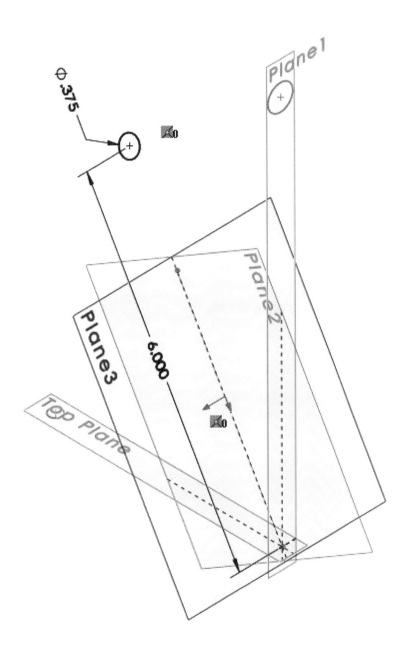

7) **Loft**  the three circles. Make sure that when you choose the circles you choose them in approximately the same location (e.g., all at the bottom quadrant or all at the top quadrant). The green circles indicate the twist. If they are all in the same location, there is no twist. You can click and drag on the green circles to adjust the twist.

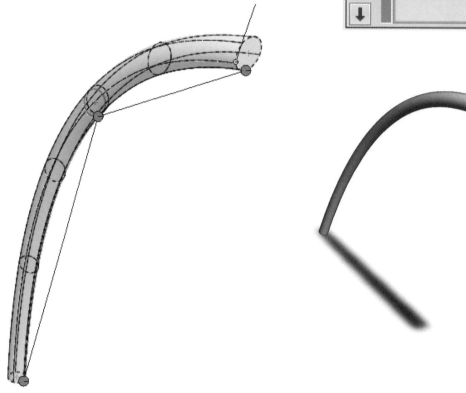

8) **Hide** ⬛ all the planes and the centerline.

9.4.5) Shell the arm

1) **Shell**  the arm to a thickness of **0.065 in** removing the small end.

2) **Sketch** and **Dimension** the following **Arcs** 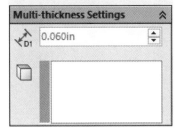 and **Centerline** on the large end face of the arm. Note that the arcs are semi-circles and are **Concentric** with the circular end. The *circle* command can also be used in conjunction with the *trim* command.

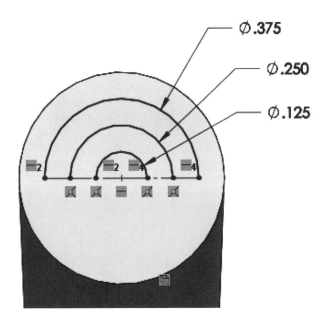

3) **Extrude Cut** 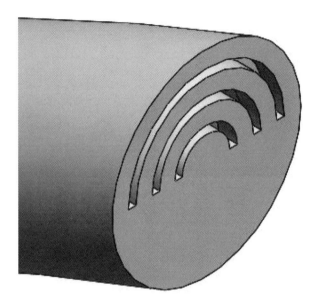 the sketch using the **Thin Feature** option with a thickness of **0.02 in** and a depth of **0.065 in.**.

9.4.6) The arm joint

1) **Sketch** and **Dimension** the following **Rectangle** on **Right Plane**.

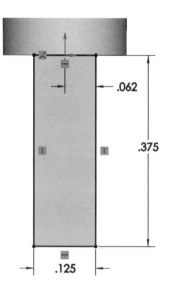

2) **Extrude**  the sketch to a distance of **0.1 in** in **both directions**.

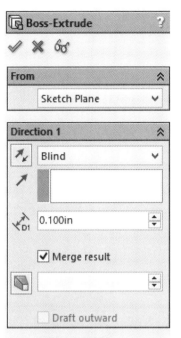

3) **Sketch** and **Dimension** the following profile on one of the large faces of the rectangular extrude.   Then, **Extrude Cut** 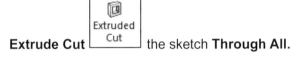 the sketch **Through All.**

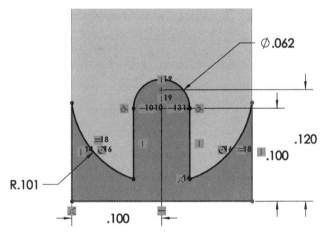

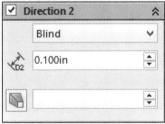

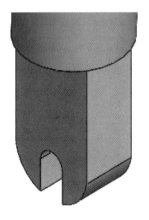

4) **Sketch** and **Dimension** the following **Circle** on the **Right Plane**. The circle center is vertical with the origin. Then, **Extrude Cut** 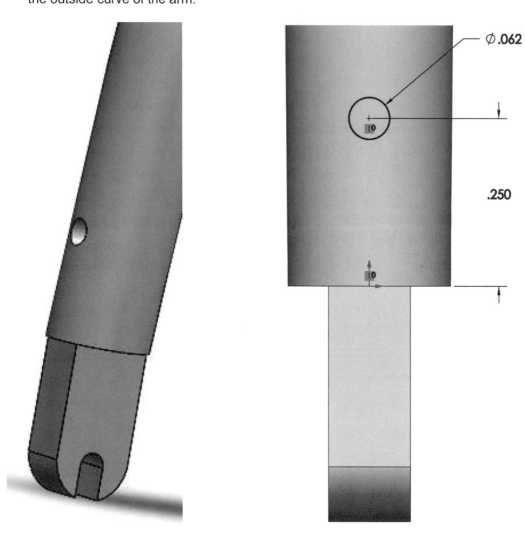 the sketch **Through All** towards the outside curve of the arm.

Ø.062

.250

5) Make the *Arm* out of **PVC Rigid**.

6) **IMPORTANT!! Save** the part and **keep** it. It will be used in a future tutorial.

9.5) BOAT TUTORIAL

9.5.1) Prerequisites

Before starting this tutorial, you should have completed the following tutorial.

- Microphone Base Tutorial

9.5.2) What you will learn

The objective of this tutorial is to introduce you to creating complex solid geometries. You will be modeling the boat shown in Figure 9.5-1. Specifically, you will be learning the following commands and concepts.

Sketching

- Ellipse
- Scale
- Construction geometry

Features

- Sweep

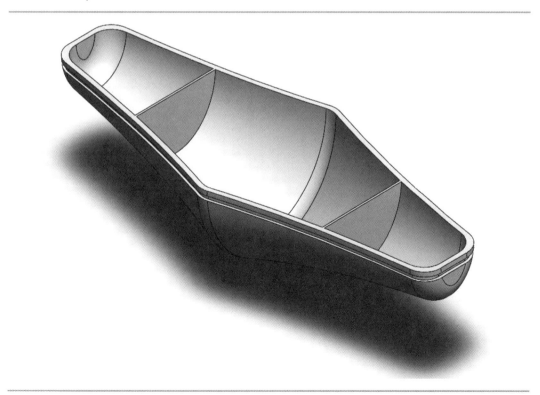

Figure 9.5-1: Boat model

9.5.3) Setting up the project

1) Start a **new part** .

2) Set the units to **MMGS,** your **Decimals = none** and the standard to **ANSI.**

3) Save the part as **BOAT.SLDPRT** (**File – Save**). Remember to save often throughout this project.

9.5.4) Loft

1) **Sketch** and **Dimension** an **Ellipse** ⊘ Ellipse on the **Right Plane**. Then draw a **Line** that cuts the ellipse in half.

> ➤ See section 9.1.3 to learn about **Ellipses** and **Conics**.

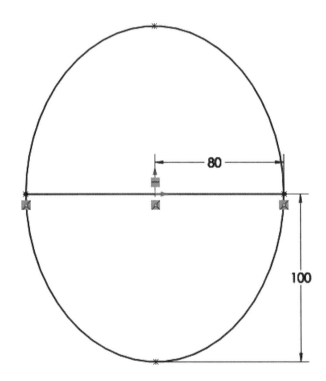

2) **Trim** 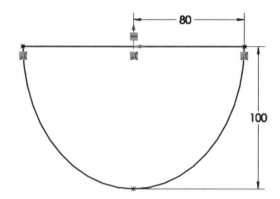 the top part of the ellipse and then **Exit Sketch.**

3) Create **4** new **Planes**  that are spaced **55 mm** apart using the **Right Plane** as the first reference.

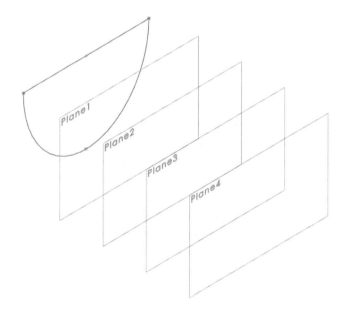

4) **Sketch** on the plane that is closest to the *Right Plane*. Use **Convert Entities** 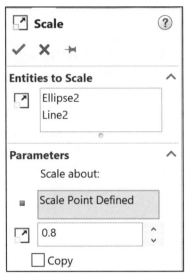 to create a new sketch on the plane that contains the half ellipse and line that was drawn on the right plane.

Scale (under the *Move Entities* command) the sketch by a factor of **80%** using the **ellipse center point** as the point to scale about. Repeat for all the planes on the right side to achieve the following sketches. Each successive sketch is 80% the size of the previous sketch.

> ➤ See section 9.1.6 to learn about *Scaling*.

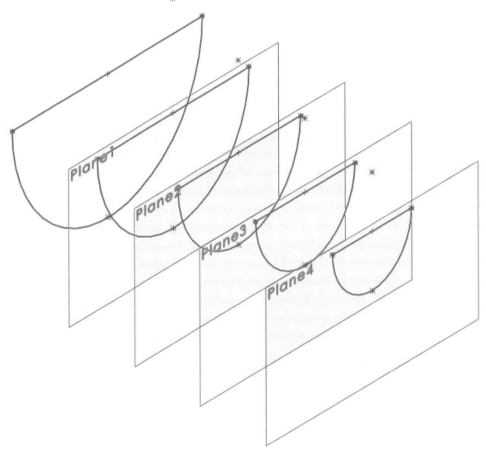

5) Use **Loft Boss/Base**  to create the right side of the boat. Select the sketches in order from largest to smallest. It is easiest if the sketches are selected in the *Design Tree*. Doing it this way avoids twist.

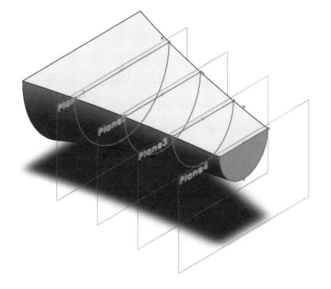

6) **Mirror** the boat using the large end as the mirror face.

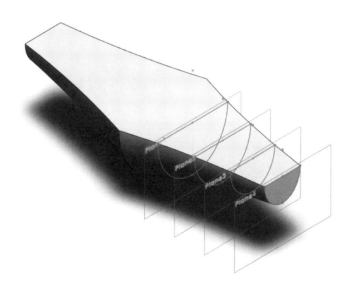

7) Hide the 4 created planes. Do this by clicking on the plane and selecting **Hide** .

9.5.5) Shell, sweep and rib

1) **Shell**  the boat with a thickness of **5 mm**. Remove the top surface.

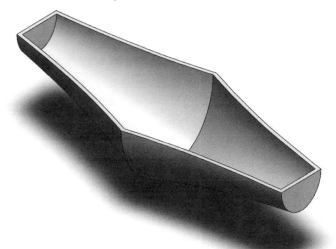

2) Apply **20 mm Fillets** to the inside and outside corners of the boat.

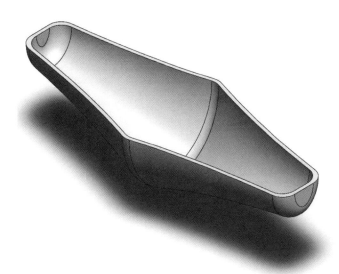

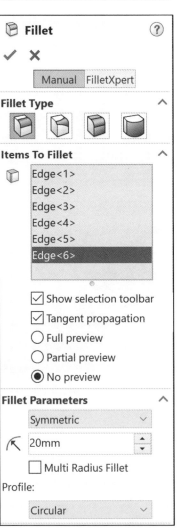

3) **Sketch** and **Dimension** the following profile on the **Right Plane** and then **Exit Sketch**.

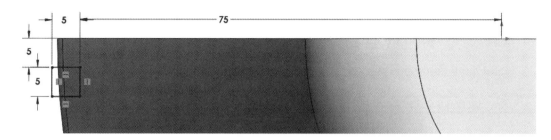

4) **Sketch** on the **Top Plane** and use **Convert Entities** [⬒ Convert Entities] to create a sketch of the outer profile of the boat. The 2 sketches will look like the second figure with the boat hidden.

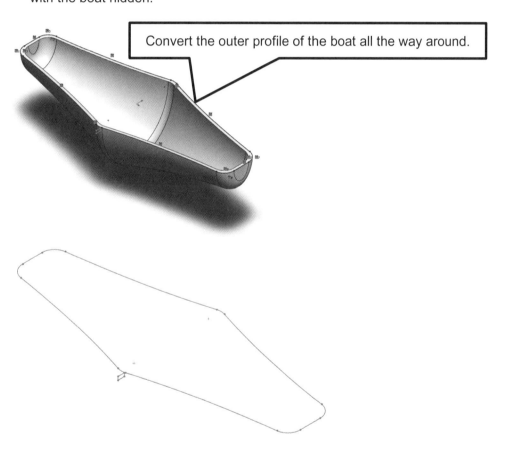

Convert the outer profile of the boat all the way around.

5) Use **Swept Cut** 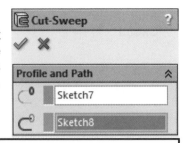 to create a groove that travels all around the boat. Use the rectangle as the profile and the converted edges as the path. It is easiest if the sketches are selected in the *Design Tree*.

Note: if the command is gray, select **Rebuild** 🔳.

> ➢ See section 9.2.2 to learn about *Sweeps*.

6) **Sketch** on **Plane 2** (the second plane created). Use **Intersection Curve**

🔲 Intersection Curve (under the *Convert Entities* command) to create the 2 lines shown. Do this by selecting the top surface of the boat. (Note: You may have to select several different locations before the top is actually selected.) Convert these lines to **Construction**

Geometry 🔲.

7) Draw a **Line** between the 2 short lines and then **Exit Sketch**.

8) **Rib** 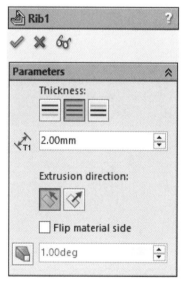 the line with **2 mm** of material on both sides. Then **Mirror** the **Rib** using the **Right Plane** as the mirror face.

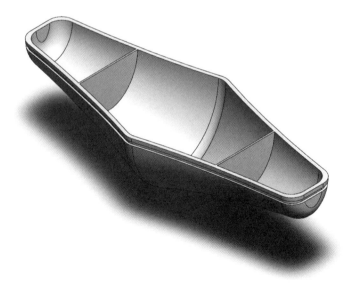

9) Make the boat out of OAK.

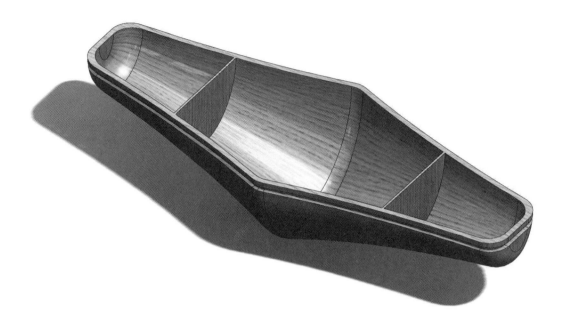

ADVANCED PART MODELING IN SOLIDWORKS® PROBLEMS

P9-1) Use SOLIDWORKS® to create a solid model of the following Cast Iron object. Use at least one LOFT command.

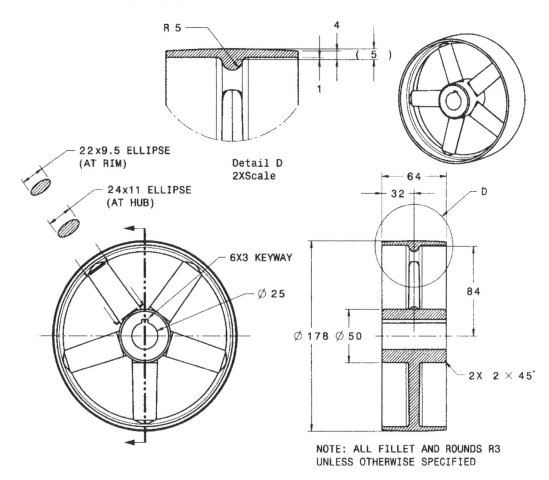

P9-2) Use SOLIDWORKS® to create a solid model of the following Cast Iron object. Use at least one RIB command.

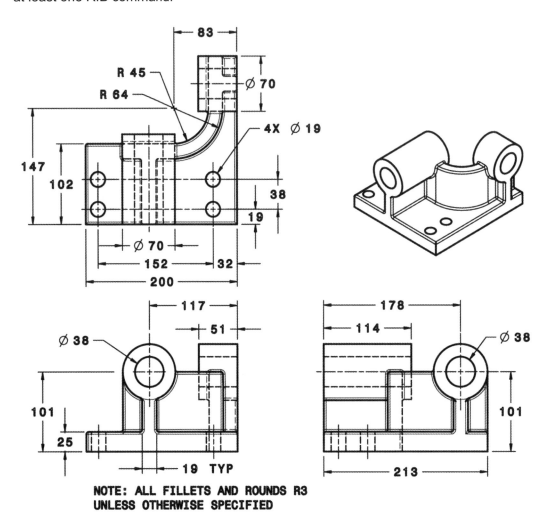

NOTE: ALL FILLETS AND ROUNDS R3
UNLESS OTHERWISE SPECIFIED

P9-3) Use SOLIDWORKS® to create a solid model of one of the following objects. Use at least one LOFT command. Apply the appropriate materials.

Wine glass	Water bottle
Screwdriver	Vase

P9-4) Use SOLIDWORKS® to create a solid model of one of the following objects. Use at least one SWEPT command. Apply the appropriate materials.

Hanger	Split cotter pin
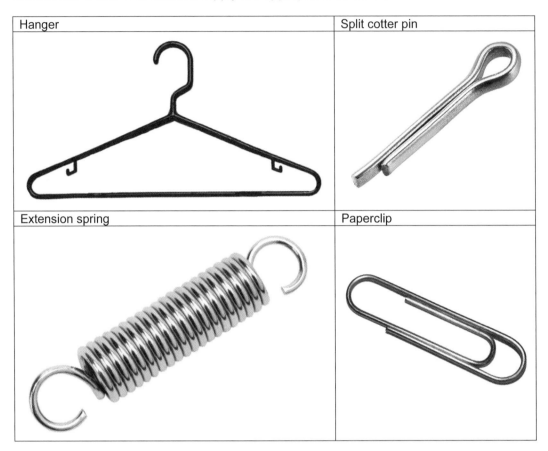	
Extension spring	Paperclip

CHAPTER 10

INTERMEDIATE ASSEMBLIES IN SOLIDWORKS®

CHAPTER OUTLINE

10.1) MATES **2**
    10.1.1) The Mate command 2
    10.1.2) Advanced Mates 2
10.2) INTERFERENCE DETECTION **4**
10.3) SENSORS **5**
10.4) ASSEMBLY VISUALIZATION **6**
10.5) TABLETOP MIRROR ASSEMBLY TUTORIAL **7**
    10.5.1) Prerequisites 7
    10.5.2) What you will learn 7
    10.5.3) The Mirror 8
    10.5.4) Setting up the assembly 9
    10.5.5) Applying mates 10
10.6) MICROPHONE ASSEMBLY TUTORIAL **12**
    10.6.1) Prerequisites 12
    10.6.2) What you will learn 12
    10.6.3) Setting up the assembly 13
    10.6.4) Applying mates 14
10.7) LINEAR BEARING TUTORIAL **16**
    10.7.1) Prerequisites 16
    10.7.2) What you will learn 16
    10.7.3) Standard mates 17
    10.7.4) Advanced mates 19
    10.7.5) Interference Detection 21
    10.7.6) Sensors and Visualization 22
INTERMEDIATE ASSEMBLIES IN SOLIDWORKS® PROBLEMS **25**

CHAPTER SUMMARY

In this chapter you will learn how to create realistic assemblies in SOLIDWORKS®. SOLIDWORKS® has many different types of mates that can be used to assemble parts. This chapter will focus on both standard and advanced mates. By the end of this chapter, you will be able to create assemblies with standard and advanced mates to produce assemblies that behave in a realistic manner.

10.1) MATES

Mates allow you to create physical constraints between parts in an assembly. For example, making surface contacts or making two shafts run along the same axis. There are three categories of mates: *Standard, Advanced, and Mechanical*.

10.1.1) The Mate command

The **Mate** command is located in the *Assembly* tab. Figure 10.1-1 shows the *Mate* options window and the available mates. Chapter 7 went over *Standard* mates and how to apply them. This chapter will focus on *Advanced* mates. There are some differences between applying a *Standard mate* and an *Advanced mate*. For the most part the steps are the same; however, some *Advanced mates* require up to four surfaces be chosen and other mates require that you fill in several parameters.

10.1.2) Advanced Mates

Along with the *Standard Mates*, there are a set of *Advanced Mates*. The available *Advanced Mates* are shown in Figure 10.1-1 and described below.

- **Profile Center:** This mate aligns two geometric profiles with their centers.

- **Symmetric:** This mate forces two similar entities to be symmetric about a plane or face of an assembly.

- **Width:** This mate centers a tab with the width of a groove.

- **Path Mate:** This mate constrains a selected point on a component to a path. Pitch, yaw, and roll of the component can be defined as it travels along the path.

- **Linear/Linear Coupler:** This mate establishes a relationship between the translation of one component and the translation of another component.

- **Distance Limit:** This mate allows a component to move within a range of distance values. A starting distance may be specified as well as a maximum and minimum value.

- **Angle Limit:** This mate allows a component to move within a range of angle values. A starting angle may be specified as well as a maximum and minimum value.

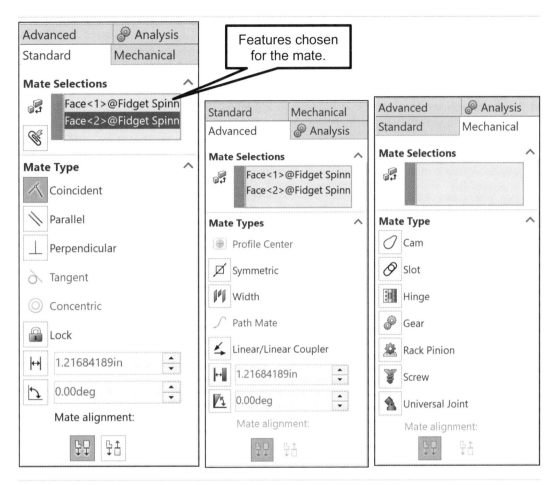

Figure 10.1-1: *Mate* Options Window

10.2) INTERFERENCE DETECTION

The **Interference detection** 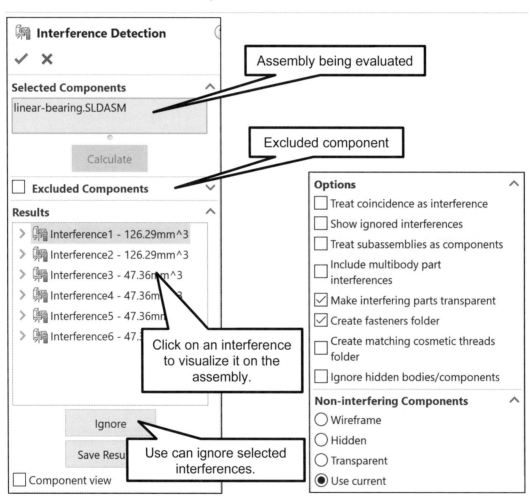 command is located in the *Evaluate* tab in the *Assembly* workspace. This command identifies interferences between parts, and helps you locate where the interferences exist. You can exclude parts that you don't want to include in the calculations. You can also ignore interferences that you want to have. For example, any press fits that exist within the assembly can be ignored. Figure 10.2-1 shows the *Interference Detection* options window.

Figure 10.2-1: Interference *Detection* options window

10.3) SENSORS

The **Sensors** [Sensor] command is located in the *Evaluate* tab in the *Assembly* workspace. *Sensors* monitor selected properties of parts and assemblies and alert you if values deviate from the limits that you have specified. Alerts will be sounded if the monitored value is

- Greater than
- Less than
- Exactly
- Not greater than
- Not less than
- Not exactly
- Between
- Not between

In some cases,

- True or False

The following properties can be monitored with a sensor.

- **Mass properties:** May be set to monitor *mass*, *volume*, and/or *surface area*.
- **Dimension:** Monitors selected dimensions.
- **Interference detection:** Monitors, in assemblies, interferences. Which means that two components physically overlap.
- **Proximity:** Monitors, in assemblies, interference between a line that is defined and a component(s) that is selected.

Sensors can also be set up to monitor results being calculated by the motion study or measured in a simulation.

10.4) ASSEMBLY VISUALIZATION

The **Assembly Visualization** command is located in the *Evaluate* tab of the *Assembly* workspace. *Assembly Visualization* provides different ways to display and sort assembly components. It sorts the components in a list, and graphically by coloring the components in the assembly. It can be sorted using the listed properties. However, there are many other properties that may be used to sort that are not listed. Figure 10.4-1 shows a *Trolley* assembly sorted by mass. The colors of the assembly correspond to the color key shown on the left.

- Mass
- Density
- Volume
- Surface area
- Face count
- Fully mated
- Quantity

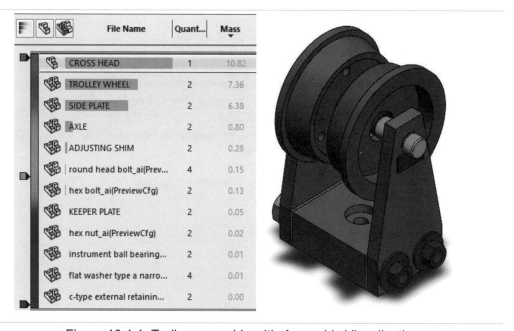

Figure 10.4-1: Trolley assembly with *Assembly Visualization*

10.5) TABLETOP MIRROR ASSEMBLY TUTORIAL

10.5.1) Prerequisites

Before starting this tutorial, you should have completed the following tutorials.

- Chapter 3 - Tabletop Mirror Tutorial
- Chapter 7 - Flanged Coupling Assembly Tutorial

10.5.2) What you will learn

The objective of this tutorial is to introduce you to using *Advanced* mates. You will be modeling the *Tabletop Mirror* assembly shown in Figure 10.5-1. Specifically, you will be learning the following commands and concepts.

Assembly

- Advanced mates
- Limit Angle mate

Figure 10.5-1: *Tabletop Mirror* Assembly

10.5.3) The Mirror

1) **Start SOLIDWORKS** and start a **new part** .

2) Set the drafting standard to **ANSI** and set the text to **Upper case for notes, tables, and dimensions**. (*Options* ⚙ *– Document Properties – Drafting Standard*)

3) Set the units to **IPS** (inch, pound, second) and the **decimal = 0.123.** (*Options* ⚙ *– Document Properties – Units*)

4) Save the part as **MIRROR.SLDPRT** in the same folder as you have the *Mirror Face* and the *Mirror Base*.

5) **Sketch** and **Dimension** the following **Circle** on the **Front Plane**.

6) **Extrude** the circle **0.25** inches.

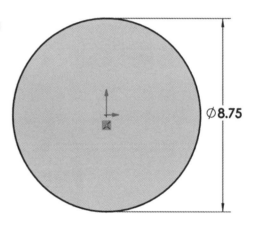

7) Make the mirror **Glass**.

8) **Save** your part.

10.5.4) Setting up the assembly

1) Open **MIRROR BASE.SLDPRT, MIRROR FACE.SLDPRT**, and **MIRROR.SLDPRT**. (Modeled in Chapter 3)

2) Create a **new Assembly** .

> ➤ See section 7.1.1 for information on starting a **New Assembly**.

3) In the *Begin Assembly* window you should see all three of your parts in the *Open documents* field. Select all the parts using either the *Shift* key or *Ctrl* key.

4) Move the mouse onto the drawing area. One of the parts will appear. Place the part by clicking the left mouse button. Another one of the parts will appear. Place this one and the next.

5) Note that all three parts appear in your *Feature Manager Design Tree.*

6) Make sure the units are **IPS,** the standard is set to **ANSI**, and then save the assembly as **TABLETOP MIRROR.SLDASM**

10.5.5) Applying mates

1) In the *Feature Manager Design Tree*, one of the parts has an (f) next to it. This is the fixed part (i.e., non-movable part). *Float* that part by right-clicking on it and selecting **Float** and then *Fix* the *Mirror Base* by right-clicking on it and selecting **Fix**.

2) Assemble the *Mirror* into the *Mirror Base* using a **Concentric** and a **Coincident** **Mate** 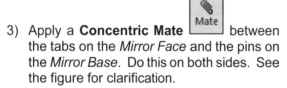.

> ➢ See section 7.2 for information on applying **Standard Mates**.

3) Apply a **Concentric Mate** 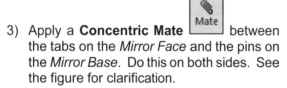 between the tabs on the *Mirror Face* and the pins on the *Mirror Base*. Do this on both sides. See the figure for clarification.

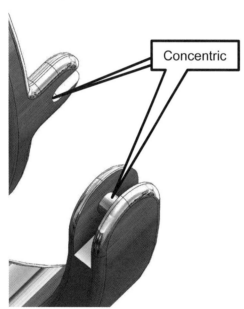

Concentric

4) **Click and drag** the *Mirror Face*. Notice that it can move back and forth. We need to fix that with the next mate.

5) Apply a **Coincident Mate** between the **Right Plane** of the Mirror *base* and the **Right Plane** of the *Mirror Face*.

6) **Click** on the *Mirror Face* and rotate it around. Notice that it will **rotate** right through the *Mirror Base*. This is not realistic. We will fix this with the next mate.

7) Apply a **Limit Angle Mate** to the *Mirror Face* to create realistic motion.

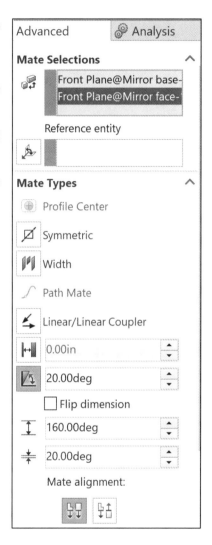

a) Select **Mate** if you are not already in the mate command.
b) Select the **Advanced** tab.
c) Select the **Front Plane** of the *Mirror Face* and the **Front Plane** of the *Mirror Base*.

d) Select the **Limit Angle** mate and enter the following parameters.
   i.   Default angle = **20** degrees
   ii. Maximum angle = **160** degrees
   iii. Minimum angle = **20** degrees

e)

> ➤ See section 10.1.2 for information on **Advanced Mates**.

8) Rotate the mirror to see that it acts in a more realistic way.

10.6) MICROPHONE ASSEMBLY TUTORIAL

10.6.1) Prerequisites

Before starting this tutorial, you should have completed the following tutorial.

- Chapter 9 - Microphone Base Tutorial
- Chapter 9 - Microphone Arm Tutorial
- Chapter 7 - Flanged Coupling Assembly Tutorial

10.6.2) What you will learn

The objective of this tutorial is to introduce you to using *Advanced* mates. You will be modeling the *Microphone* assembly shown in Figure 10.6-1. Specifically, you will be learning the following commands and concepts.

Assembly

- Advanced mates
- Limit Angle mate

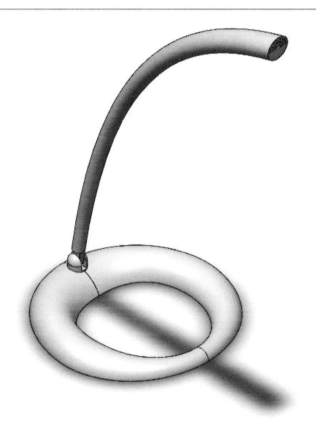

Figure 10.6-1: *Microphone* Assembly

10.6.3) Setting up the assembly

1) Open **MICROPHONE BASE.SLDPRT** and **MICROPHONE ARM.SLDPRT**

2) Open a **New** ▭ **Assembly** .

3) From the *Begin Assembly* window, insert the **MICROPHONE BASE** and **MICROPHONE ARM**. Select both file names and click in the drawing area to place the components.

4) Make sure the units are **IPS** and the standard is **ANSI**. Save the assembly as **MICROPHONE.SLDASM**. At completion of the tutorial, the assembly's *Feature Design Tree* will look like what is shown.

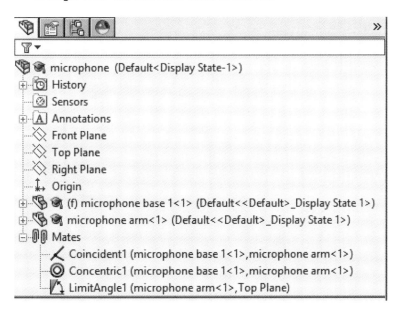

10.6.4) Applying mates

1) Make sure that the *MICROPHONE BASE* is fixed, and the *MICROPHONE ARM* is floating. To do this, right click on the part and either select **Fix** or **Float**. Notice that the *base* will have an (f) next to it indicating that it is fixed.

2) Apply the following **Mates** .
   a. A **Coincident** mate between the **Right Plane** of the *BASE* and the **Front Plane** of the *ARM*.

      Reverse the **Mate alignment** if necessary.
   b. A **Concentric** mate between the round shaft of the *base joint* and the round slot of the *arm joint*.

   ➤ See section 7.2 for information on applying ***Standard Mates***.

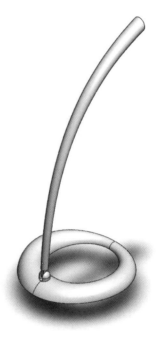

3) Click and drag the microphone arm to see its motion. Note that it will pass through the base, which is not very realistic.

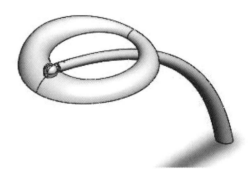

4) Apply A **Limit Angle Mate** to produce a realistic motion of the *Arm*.
   a. Move the *ARM* so that it makes an acute angle with the *BASE*.
   b. Select **Mate** 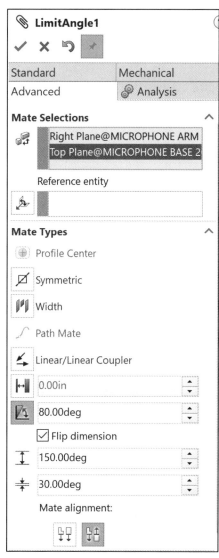 if you are not already in the mate command.
   c. Select the ***Advanced*** tab.
   d. Select the **Right Plane** of the *Arm* and the **Top Plane** of the *Base*.
   e. Select the **Limit Angle** mate and enter the following parameters.
      i. Default angle = **80** degrees
      ii. Maximum angle = **150** degrees
      iii. Minimum angle = **30** degrees
   f.

> See section 10.1.2 for information on **Advanced Mates**.

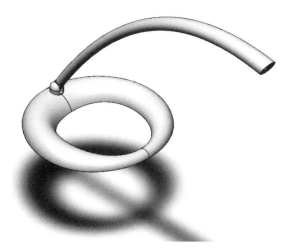

5) Click and drag the microphone arm to see its motion. Note that it does not pass through the base now.

10.7) LINEAR BEARING TUTORIAL

10.7.1) Prerequisites

Before starting this tutorial, you should have completed the following tutorial.

- Chapter 7 - Flanged Coupling Assembly Tutorial

10.7.2) What you will learn

The objective of this tutorial is to introduce you to *Advanced Mates*. You will assemble the parts that comprise the *Linear Bearing* assembly shown in Figure 10.7-1. Specifically, you will be learning the following commands and concepts.

Assembly

- Advanced mates
- Distance mate
- Width mate

Evaluate

- Interference Detection
- Sensors
- Assembly Visualization

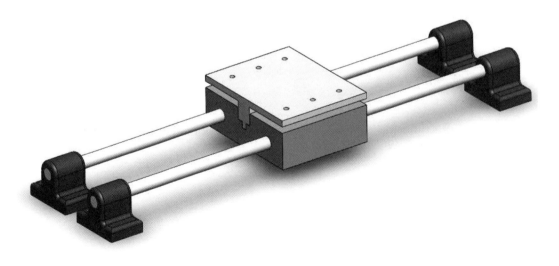

Figure 10.7-1: Linear bearing assembly

10.7.3) Standard mates

1) Download the following files and place them in a common folder.
   - **linear-bearing-student.SLDASM**
   - **attachment-plate.SLDPRT**
   - **bearing-support.SLDPRT**
   - **guide-bar.SLDPRT**
   - **guide-block.SLDPRT**

2) Open **linear-bearing-student.SLDASM**. Notice which components are fixed and which are floating. A fixed component will have an **(f)** in front of its name. The four bearing supports should be fixed. To change a component from fixed to float, or the other way around, right click on the component and choose either **fix** or **float**.

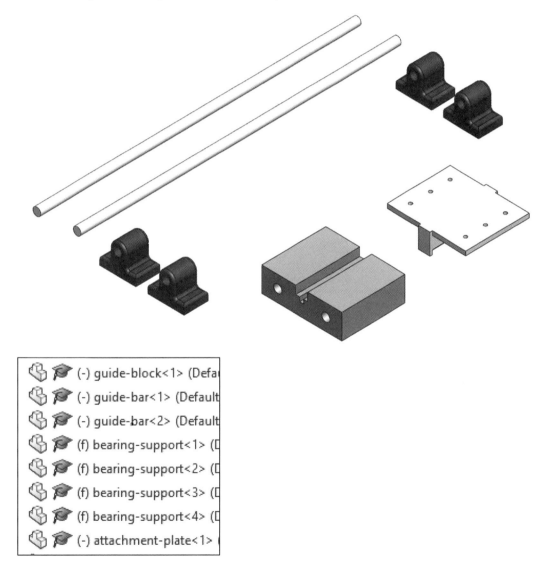

3) Apply the following **Standard mates**.
   - **Concentric** between the **Guide Bars** and the appropriate **Bearing Support**.
   - Make one end of each **Guide Bar Coincident** with the face of the **Bearing Support**.
   - **Concentric** between the **Guide Bars** and the **Guide Block**.

> ➤ See section 7.2 for information on applying **Standard Mates**.

Once completed, your linear bearing assembly should look like the figure, and you should have a total of 6 active mates. <u>Note:</u> There will be some greyed out mates that I used to position the *Bearing Supports* initially that are now inactive.

◎ Concentric1 (guide-bar<2>,bearing-support<4>)

◎ Concentric2 (guide-bar<1>,bearing-support<3>)

⋏ Coincident17 (guide-bar<1>,bearing-support<2>)

⋏ Coincident18 (guide-bar<2>,bearing-support<1>)

◎ Concentric3 (guide-block<1>,guide-bar<1>)

◎ Concentric4 (guide-block<1>,guide-bar<2>)

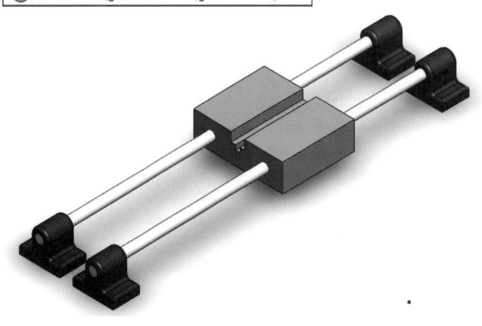

10.7.4) Advanced mates

1) Move the *Guide Block* to the left and notice that it will pass through the *Bearing Supports*. This is not realistic. Move the *Guide Block* back to the center.

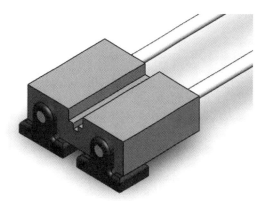

> ➤ See section 10.1.2 for information on **Advanced Mates**.

2) Apply a **Limit Distance mate** between the **Guide Block** and the **Bearing Support** to make it move more realistically.

a) Select **Mate** 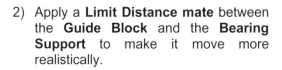 if you are not already in the mate command.
b) Select the **Advanced** tab.
c) Select the **Face 1** of the *Guide Block* and then **Face 2** of the *Bearing Support*.

d) Select the **Limit Distance** mate and enter the following parameters.
   i. Default distance = **200** mm
   ii. Maximum distance = **260** mm
   iii. Minimum angle = **0**
e)
f) Test the mate to see if it works properly.

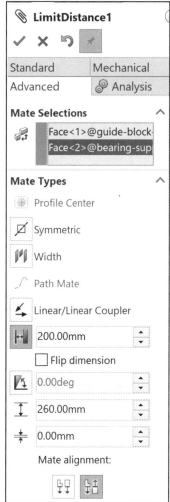

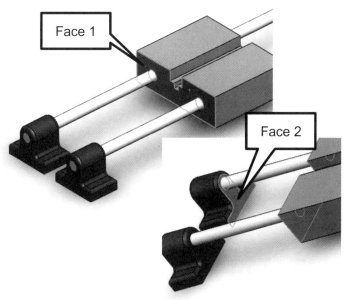

3) Apply the following **Width Mate** to center the *Attachment Plate* in the *Guide Block*.

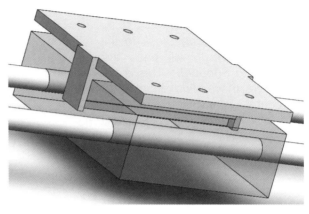

a) Select **Mate** 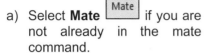 if you are not already in the mate command.

b) Select the ***Advanced*** tab.

c) Select the **Width** mate

d) Width Selection: Select the two inside surfaces of the ***Guide Block*** indicated in the figure.

e) Tab selection: Select the two outside surfaces of the ***Attachment Plate*** indicated in the figure.

f) ☑

4) Apply a **Coincident** mate between the bottom surface of the ***Guide Block*** groove and bottom surface of the ***Attachment Plate*** tab.

5) Apply a **Coincident** mate between the end of the ***Guide Block*** and inner surface of the ***Attachment Plate*** groove (see figure).

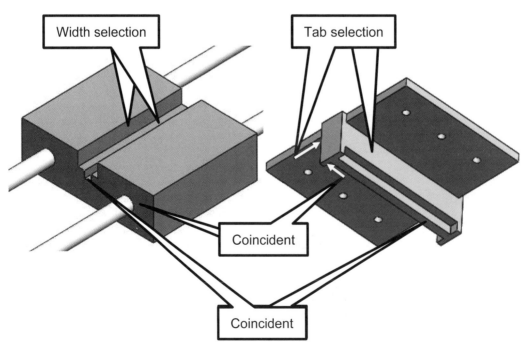

Width1 (Guide block<1> ,Attachment plate<1>)
Coincident21 (Guide block<1>,Attachment plate<1>)
Coincident22 (Guide block<1>,Attachment plate<1>)

Width selection

Tab selection

Coincident

Coincident

10.7.5) Interference Detection

1) Check for interferences in the assembly or parts that overlap.

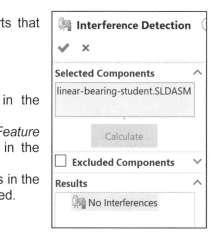

a) Select **Interference Detection** in the *Evaluate* tab.
b) Select **linear-bearing assembly** in the *Feature Design Tree* if it does not already appear in the *Selected Components* area.
c) Click on **Calculate** to detect any interferences in the assembly. No interferences should be detected.
d)

> ➢ See section 10.2 for information on ***Interference Detection***.

---

2) Create an interference by changing the diameter of the *Guide bar* to **10.1 mm**. Note that the holes in the *Bearing Supports* and *Guide Block* are 10 mm.
   a) **Right-click** on one of the *Guide bars* in the *Feature Design Tree* and select **Edit Part** 🔧.
   b) Change the diameter of the sketch used in the *Extrude* from 9.8 to **10.1 mm**.
   c) **Right-click** on one of the *Guide bars* in the *Feature Design Tree* and select **Edit Assembly**.
   d) **Rebuild** the assembly.
   e) Run the **Interference Detection** again. Several interferences should be detected. When you click on each interference, it should be highlighted in red.
   f)

3) Change the diameter of the *Guide bar* back to **9.8 mm**.

10.7.6) Sensors and Visualization

1) Add a **Sensor** to detect when the design exceeds a mass of **2000 g**.

   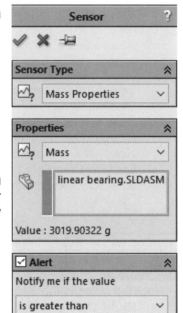

   a) Select **Sensor**  in the *Evaluate* tab.
   b) Sensor type = Mass Properties
   c) Properties = Mass
   d) Entities to Monitor = Linear Bearing Assembly
   e) **Check** the **Alert** checkbox.
   f) Select **'is greater than'** and enter **2000**.

   g) ![check]. Note that when you apply the sensor, you automatically get an alert. This is because your assembly mass is 3020 g. Sensors are usually applied before you start designing.

   > ▸ 🕮 ⚠ Sensors
   >   ⚖ ⚠ Mass1(3049.16 g)

   > ➢ See section 10.3 for information on **Sensors**.

2) Use **Assembly Visualization** to colorize your assembly based on the **mass** of each part.

   a) Select **Assembly Visualization** 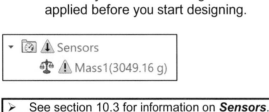 in the *Evaluate* tab.
   b) Mass should be the default visualization parameter. Click the arrow that points to the right and select **Mass**. This lists all the properties that you can use to visualize your assembly. Notice that the *Guide Block* is the heaviest and that the assembly is now color coded. If your assembly is not colored, click on key that indicates color.

   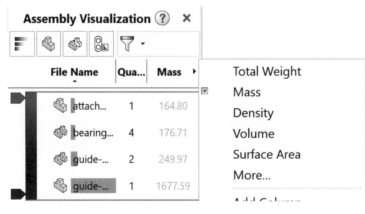

   c) Colors in the legend may be added or taken away and the sliders may be moved. Practice with the colors.

   > ➢ See section 10.4 for information on **Assembly Visualization**.

3) Use **Surface Area** as the visualization parameter and add a mid-tone color by clicking to the left of the color bar. Click on **Surface area** to arrange from largest to smallest.

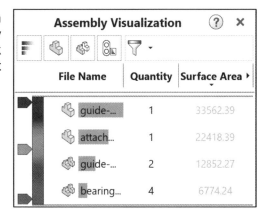

4) Click on the **Group/Ungroup View** 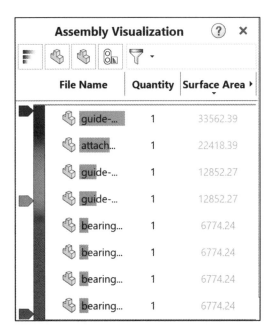 icon. This will show every individual part and not group like parts.

5) **Save**.

NOTES:

INTERMEDIATE ASSEMBLIES IN SOLIDWORKS® PROBLEMS

P10-1) Model one of the following. Use the appropriate materials and mates.

a) Scissors

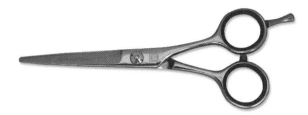

b) Pliers

c) Carabiner

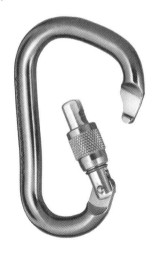

NOTE:

CHAPTER 11

TOLERANCING AND THREADS IN SOLIDWORKS®

CHAPTER OUTLINE

11.1) THREADS...2
    11.1.1) What is a thread?...2
    11.1.2) How are threads represented on a drawing? 2
    11.1.3) What is a thread note?...3
11.2) MODELING THREADS 5
    11.2.1) Thread Command 5
    11.2.2) Cosmetic Threads...7
    11.2.3) Stud Wizard 8
11.3) TOLERANCES 10
    11.3.1) What is a tolerance? 10
    11.3.2) Types of tolerances 11
11.4) APPLYING TOLERANCES..11
11.5) VISE - STATIONARY JAW TUTORIAL...15
    11.5.1) Prerequisites 15
    11.5.2) What you will learn..15
    11.5.3) Setting up the part 17
    11.5.4) Modeling the part 17
    11.5.5) Applying threads 21
    11.5.6) Applying tolerances 24
11.6) VISE - SCREW TUTORIAL 25
    11.6.1) Prerequisites 25
    11.6.2) What you will learn..25
    11.6.3) Setting up the part 26
    11.6.4) Modeling the part 26
    11.6.5) Applying threads 28
    11.6.6) Applying Tolerances 29
TOLERANCING AND THREADS IN SOLIDWORKS® PROBLEMS............................31

CHAPTER SUMMARY

In this chapter, you will learn about modeling threads and creating toleranced parts. If a feature's size is toleranced, it is allowed to vary within a range of values or limits. It is no longer controlled by a single size. By the end of this chapter, you will be able to apply tolerances to a basic dimension and represent threads realistically or cosmetically.

11.1) THREADS

11.1.1) What is a thread?

A screw thread is a ridge of uniform section in the form of a helix. This allows the thread to advance when it is turned. Threads are an important joining method when you want to be able to put together and take apart the assembly. Figure 11.1-1 shows a detailed representation of an external and internal thread.

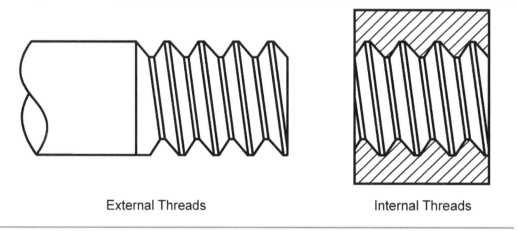

External Threads   Internal Threads

Figure 11.1-1: Detailed representation of external and internal threads.

11.1.2) How are threads represented on a drawing?

There are three methods of representing screw threads on a drawing: **detailed, schematic**, and **simplified**. There are advantages and disadvantages to using each type. Figure 11.1-1 shows the detailed representation of threads. The advantage of this type of thread representation is that it is highly effective in that it looks like a thread, but it takes a lot of computer memory. Figure 11.1-2 shows the schematic representation of threads. The advantage of this type of thread is that it is almost as effective at representing a thread and it doesn't take as much computer memory as the detailed representation. The disadvantage is that there are a lot of lines that could obscure other features of the drawing. Figure 11.1-3 shows the simplified representation of threads. This type of representation is not very effective at representing threads but doesn't use much memory. This is the default method that SOLIDWORKS® uses to represent threads on a drawing. Over time you will get used to seeing threads when looking at the simplified representation of threads.

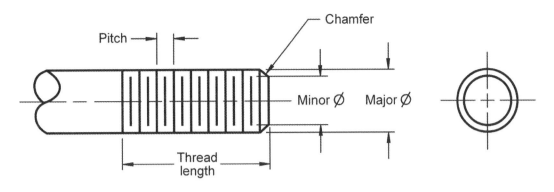

Figure 11.1-2: Schematic representation of external threads.

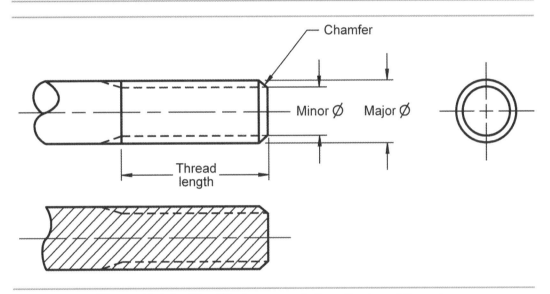

Figure 11.1-3: Simplified representation of external threads.

11.1.3) What is a thread note?

Threads need to be identified on a drawing using a thread note. Each type of thread (e.g., Unified, Metric) has its own way of being identified. Unified threads (inch) are identified in a thread note by the *major diameter*, *threads per inch*, *thread form* and *series*, *thread class*, whether the thread is *external* or *internal*, whether the thread is *right* or *left*-handed, and the thread *depth* (internal only) as shown in Figures 11.1-4.

Metric threads notes are identified in a thread note by the *metric form symbol*, *major diameter*, *pitch*, *tolerance class*, whether the thread is *right* or *left*-handed, and the thread *depth* (internal only) as shown in Figures 11.1-5.

Both the Unified and Metric thread notes are commonly shown in an abbreviated form. Usually the *thread class/tolerance class* and whether the thread is *right* or *left*-handed is left off unless these values are not standard.

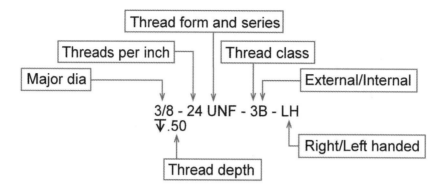

Figure 11.1-3: Unified thread note components

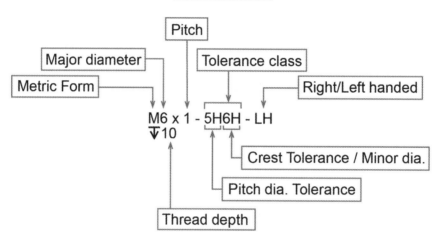

Figure 11.1-5: Metric thread note components

11.2) MODELING THREADS

11.2.1) Thread Command

The **Thread** command is located in the *Feature* tab stacked under the *Hole Wizard*. It allows you to add threads to the part automatically, without having to use a *Helix* and the *Sweep* command to create the threads. Figure 11.2-1 shows a bolt with no threads applied and a bolt after using the *Thread* command to apply threads. It also shows how that bolt will look on a drawing. Use the following steps to apply threads. Figure 11.2-2 shows the *Thread* option window.

Applying Threads

1) Select the **Thread** [Thread] command.
2) Select a circular edge where the threads will start. (See Figure 11.2-2.)
3) End condition: For the end condition you can select a thread length, number of revolutions or have the threads travel up to another surface.
4) Specifications:
   a) Type: Select the thread standard.
   b) Size: Select the thread size and series (i.e., Pitch for metric threads, threads per inch for Unified threads).
   c) Thread method: You can choose to cut the threads into the material (Cut thread) or lay the threads on top of the material (Extrude thread).
5) Thread Options: Choose whether the threads are right-handed or left-handed.

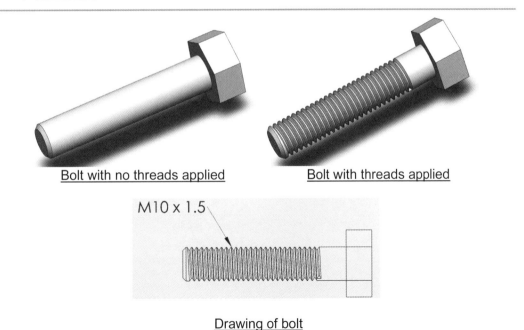

Bolt with no threads applied   Bolt with threads applied

M10 x 1.5

Drawing of bolt

Figure 11.2-1: Applying the *Thread* command to a bolt

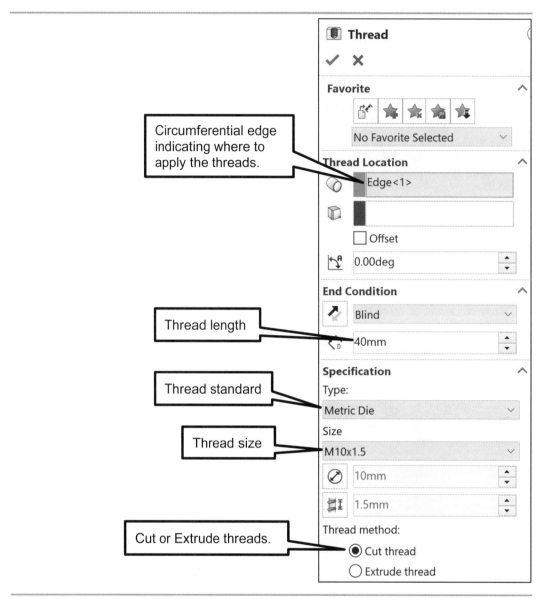

Circumferential edge indicating where to apply the threads.

Thread length

Thread standard

Thread size

Cut or Extrude threads.

Figure 11.2-2: *Thread* options window

11.2.2) Cosmetic Threads

Cosmetic Threads describe a threaded feature without having to model real threads. For example, if we want to apply threads to the shaft of a bolt shown in Figure 11.2-3, we can use *Cosmetic Threads*. After the threads are applied, a dotted circle at the end of the shaft can be seen. This indicates that the threads have been applied, and it takes a lot less computer memory to model threads in this way. *Cosmetic Threads* may also be applied to hole. Even though the threads are cosmetic, they will still show up on the drawing. The drawing will show that threads have been applied using the simplified thread symbol as shown in Figure 11.2-3.

The **Cosmetic Thread** command can be accessed through the pull-down menu. Use the following steps to apply cosmetic threads. Figure 11.2-4 shows the *Cosmetic Thread* option window.

Applying Cosmetic Threads

1) From the pull-down menu choose **Insert – Annotation – Cosmetic Thread…**
   Cosmetic Thread…
2) Select a circular edge where the threads will start. (See Figure 11.2-4.)
3) Standard: Select the thread standard.
4) Type: Select the type of thread.
5) Size: Select the thread size and series (i.e., Pitch for metric threads, threads per inch for Unified threads).
6) End condition: For the end condition a thread length, up to next, or through can be selected.
7) If needed, select the thread class.

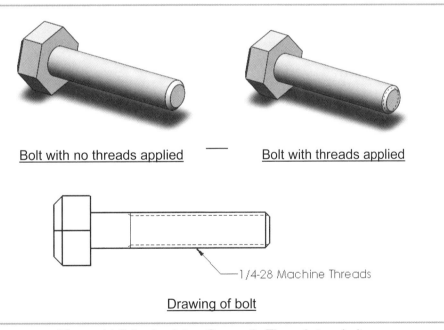

Bolt with no threads applied — Bolt with threads applied

1/4-28 Machine Threads

Drawing of bolt

Figure 11.2-3: Applying *Cosmetic Threads* to a bolt

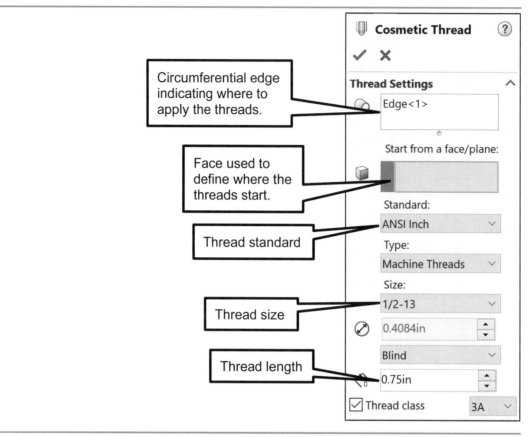

Figure 11.2-4: *Cosmetic Thread* options window

11.2.3) Stud Wizard

A stud is a piece of threaded rod. It's just like a machine screw without a traditional head. It could be just a rod with no head or have a nontraditional head. Studs are often permanently attached to a part providing a means for attaching other parts of the assembly.

The **Stud** command is located in the *Feature* tab stacked under the *Hole Wizard*. The *Stud* command is very similar to the *Thread* command. The one significant difference is the ability within the *Stud* command to add an undercut after the threads have ended.

Figure 11.2-5 shows a rod with threaded ends. The threaded ends were created using the *Stud* command. One end uses an undercut and the other does not. Notice that the *Stud* command uses cosmetic threads. This figure also shows how the drawing will look. To apply threads using the *Stud* command, use the following steps. Figure 11.2-6 shows the *Stud Wizard* option window.

Applying threads using the *Stud* command

1) Select the **Stud Wizard** | 🔩 Stud Wizard | command.
2) Select a circular edge where the threads will start. (See Figure 11.2-2)
3) <u>Standard:</u> Select the thread standard.
    a) <u>Type:</u> Select the type of thread.
    b) <u>Size:</u> Select the thread size and series (i.e., Pitch for metric threads, threads per inch for Unified threads).
4) <u>Thread Class:</u> If necessary, specify a thread class.
5) <u>Undercut:</u> If an undercut is needed, click the checkbox and fill in the dimensions.

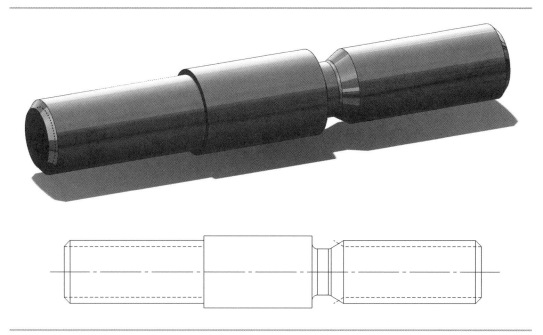

Figure 11.2-5: Applying threads using the *Stud* command

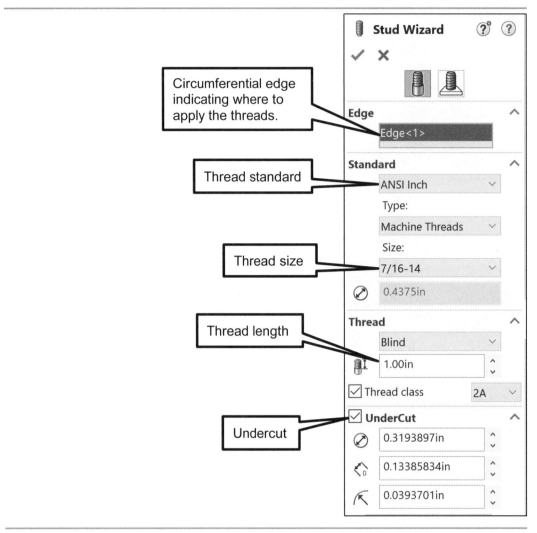

Figure 11.2-6: *Stud Wizard* options window

11.3) TOLERANCES

11.3.1) What is a tolerance?

A tolerance is the total amount that a dimension is permitted to vary. In other words, it is the difference between the maximum and minimum size of the feature. Tolerancing enables similar parts to be near enough alike so that any one of them will fit properly into the assembly. Tolerancing gives us the means of specifying dimensions with whatever degree of accuracy we may require for our design to work properly. We would like to choose a tolerance that is not unnecessarily accurate or excessively inaccurate. Choosing the correct tolerance for a particular application depends on the design intent (i.e., the end use) of the part, cost, how it is manufactured, and experience.

Standards are needed to establish dimensional limits for parts that are to be toleranced. The two most common standards agencies are the American National Standards Institute (ANSI) and the International Standards Organization (ISO).

11.3.2) Types of tolerances

Toleranced dimensions may be presented using three general methods: *limit* dimensions, *plus-minus* tolerances, and *page* or *block* tolerances. Figure 11.3-1 shows an example of all three methods. Limit and plus-minus tolerances are used for individual dimension whereas a block tolerance is a general note that applies to all dimensions that are not covered by some other tolerancing type. Block tolerances are placed in the tolerance block to the left of the title block.

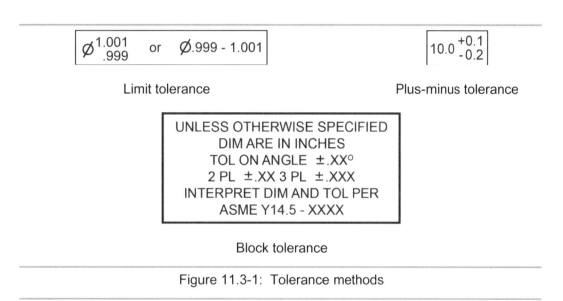

Limit tolerance   Plus-minus tolerance

Block tolerance

Figure 11.3-1: Tolerance methods

11.4) APPLYING TOLERANCES

In SOLIDWORKS®, tolerances can be applied within the *Dimension options* window. Figure 11.4-1 shows a circle with a basic dimension and that same circle with a tolerance applied. To apply a tolerance to a dimension, use the following steps. The *Dimension* option window is shown in Figure 11.4-2.

Manually applying the tolerance

1) Click on the dimension that you wish to tolerance. Note: If you click on the dimension value, you need to double-click.
2) In the *Dimension* option window, select the tolerance appearance (e.g., limit, bilateral, fit). Note that bilateral is the same as plus-minus.
3) If you chose an appearance that requires you to input your upper and lower variation, do that. If you chose an appearance that requires you to choose a shaft and hole fit, do that. (Note: You can force your lower variation to be a positive number by putting a negative in front of the negative number. That's 2 negatives.)
4) Select the number of decimal places.

You can also apply tolerances to a hole if the hole was created using the *Hole Wizard*. Use the following steps to apply a tolerance to a hole. The *Hole Specification* options window is shown in Figure 11.4-3.

Applying tolerances to a *Hole*

1) Create the **Hole** .
2) Expand the **Tolerance/Precision** area.
3) Select the tolerance appearance (e.g., limit, bilateral, fit).
4) If you chose an appearance that requires you to input the upper and lower variation, do that. If you chose an appearance that requires you to choose a shaft and hole fit, do that. (Note: You can force the lower variation to be a positive number by putting a negative in front of the negative number. That's 2 negatives.)
5) Select the number of decimal places.
6) <u>Note:</u> If you edit the sketch that was used to create the *Hole*, you will see a toleranced dimension.

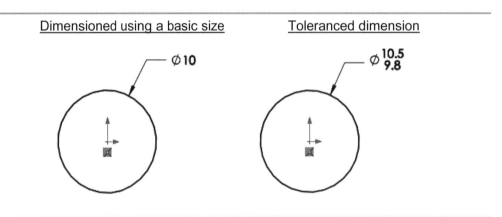

Figure 11.4-1: Toleranced dimension

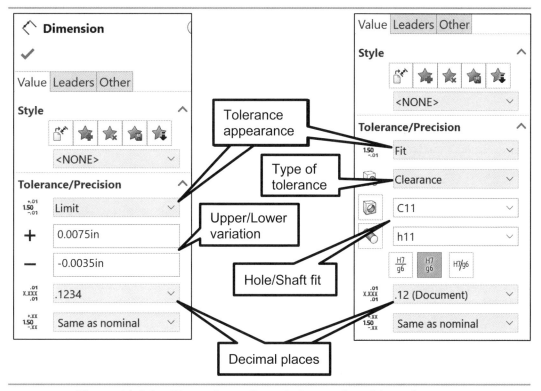

Figure 11.4-2: *Dimension* option window and applying tolerances

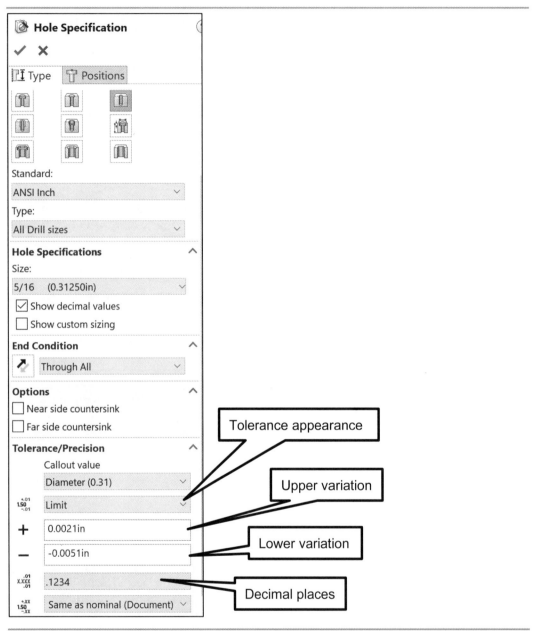

Figure 11.4-3: Tolerancing a *Hole*

11.5) VISE - STATIONARY JAW TUTORIAL

11.5.1) Prerequisites

Before starting this tutorial, you should have completed the following tutorials.

- Chapter 1 - Connecting Rod Tutorial
- Chapter 3 - Flanged Coupling Tutorial

You should also have the following knowledge.

- Familiarity with threads and fasteners
- Creating threaded holes using the *Hole Wizard*
- Familiarity with tolerancing

11.5.2) What you will learn

The objective of this tutorial is to introduce you to realistic part models that have toleranced dimensions. You will be modeling the *Stationary Jaw* from the *Vise* assembly shown in Figure 11.5-1. Specifically, you will be learning the following commands and concepts.

Sketching

- Applying tolerances

Features

- Cosmetic Thread
- Thread

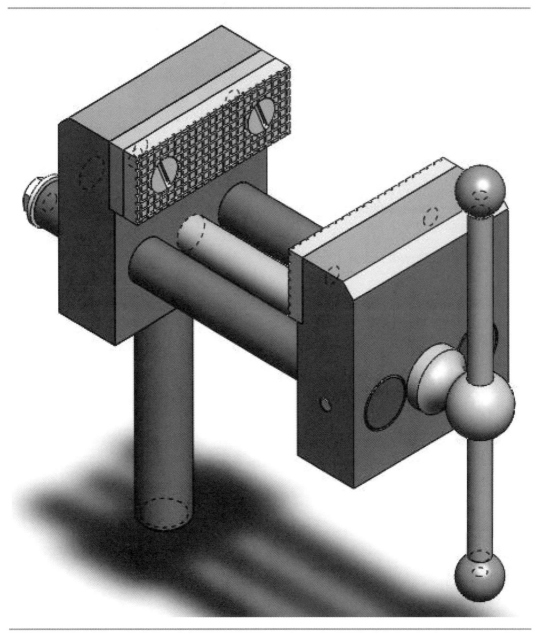

Figure 11.5-1: Vise assembly

11.5.3) Setting up the part

1) Start a **new part** and save the part as **STATIONARY JAW.SLDPRT** (**File – Save**). Remember to save often throughout this project.

2) Set the drafting standard to **ANSI** and set the text to **upper case** for notes, tables, and dimensions. (*Options* 🔘 *– Document Properties – Drafting Standard*)

3) Set the units to **IPS** (i.e. inch, pound, second) and set the **Decimals = .12**. Also, select the rounding option, **Round half to even**. (*Options* 🔘 *– Document Properties – Units*)

11.5.4) Modeling the part

1) You will be modeling the *Stationary Jaw* shown on the next page. For now, model the part without any of the holes or threads. For the threaded shaft, model it as if it were a 0.75-inch diameter rod.

2) Make the part **AISI 1020 Steel, Cold Rolled**.

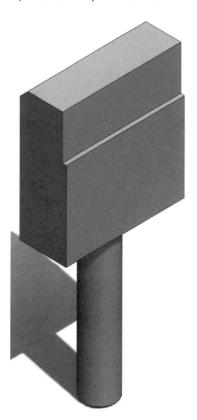

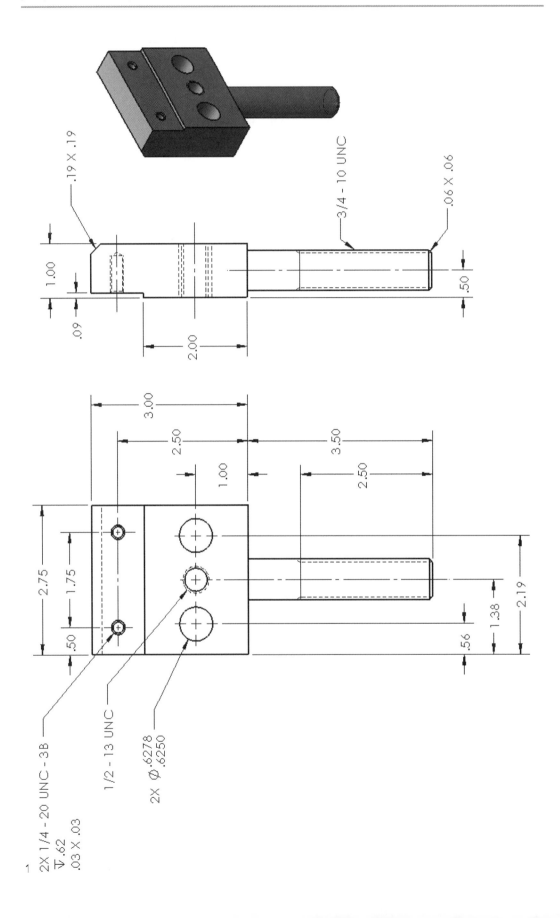

3) Use the **Hole Wizard** 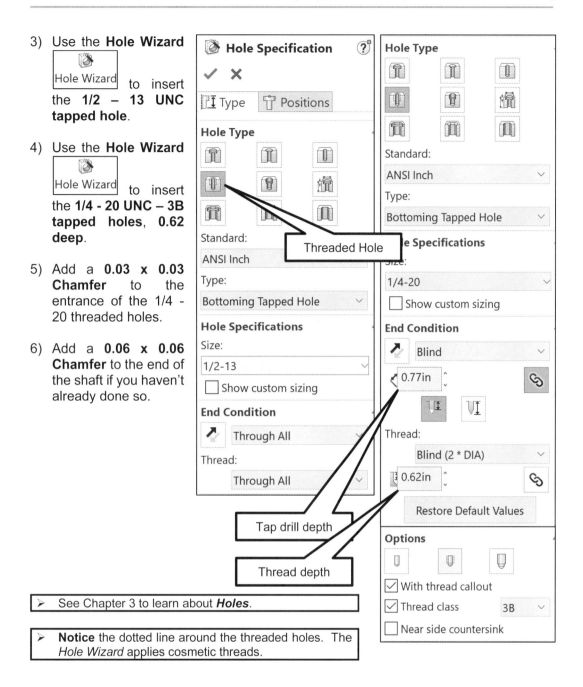 to insert the **1/2 – 13 UNC tapped hole**.

4) Use the **Hole Wizard** to insert the **1/4 - 20 UNC – 3B tapped holes, 0.62 deep**.

5) Add a **0.03 x 0.03 Chamfer** to the entrance of the 1/4 - 20 threaded holes.

6) Add a **0.06 x 0.06 Chamfer** to the end of the shaft if you haven't already done so.

➤ See Chapter 3 to learn about **Holes**.

➤ **Notice** the dotted line around the threaded holes. The *Hole Wizard* applies cosmetic threads.

7) Add the two holes that are toleranced using a basic size of **5/8 inch**. Use a **Sketch** and an **Extruded Cut**. We will tolerance these holes in a later step.

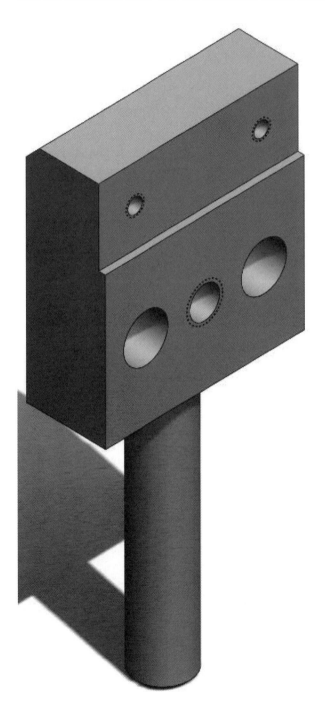

11.5.5) Applying threads

1) Use the **Thread** command to apply the 3/4 – 10 UNC threads to the shaft.

   a) Select the **Thread** [🔩 Thread] command (under the *Hole Wizard*).
   b) Thread Location:
      I.   Select a circular edge at the end of the shaft that is the same size as the major diameter of the threads.
      II. Select the box under the *Edge of Cylinder* box and select where you want your threads to start.
   c) End condition: Select **Blind** and enter **2.50** inch.
   d) Specifications:
      I.   Type: **Inch Die**
      II. Size: **0.7500 - 10**
      III. Thread method: **Cut thread**

   e) [✓]

> See section 11.1 and 11.2 to learn about *Threads*.

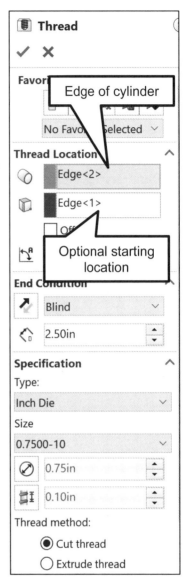

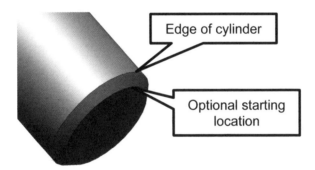

2) Notice that the threads don't cut all the way out the end. Let's fix that.
   a) **Edit** the **Thread** you just applied.
   b) Check the **Offset** check box.
   c) Enter **0.10** inch and flip the direction so the threads go off the end.
   d) Since we have shifted the threads to go 0.1 inch off the end, we need to adjust the thread length to **2.60** inch.

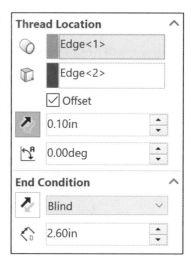

3) Suppress the *Threads* that were just applied. **Right-click** on the *Thread* in the *Feature Tree* and select **Suppress**

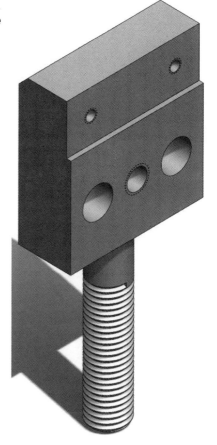

4) Apply **Cosmetic Threads** to the shaft.
   a) From the pull-down menu choose **Insert –
      Annotation – Cosmetic Thread…**

       Cosmetic Thread…

   b) Select a circular edge at the end of the shaft
      that is the same size as the major diameter of
      the threads.
   c) <u>End condition:</u> Choose **Blind** and enter **2.5** inch.
   d) Select the box under the *Edge of Cylinder* box
      and select where you want your threads to start.
      Note: This option only appears after you have
      selected the *Blind* end condition.
   e) <u>Standard:</u> **ANSI Inch**
   f) <u>Type:</u> **Machine Threads**
   g) <u>Size:</u> **3/4 – 10**
   h) Notice that a dotted line appears at the end of
      the shaft indicating that the threads have been
      applied.

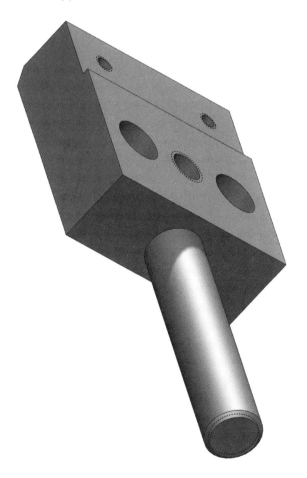

11.5.6) Applying tolerances

1) When modeling the part initially, basic sizes were used. The 2X **n**.6250 - .6278 dimension was modeled as a 5/8-inch diameter hole. We will apply the tolerances now.

> ➤ See section 11.3 and 11.4 to learn about **Tolerances**.

a) **Edit** the **Sketch** attached to the *Extruded Cut* in your *Feature Tree* that was used to create the holes.
b) **Click** on the **diameter** dimension. If you click on the dimension value, you will have to double click.
c) Select a tolerance appearance of **Limit**.
d) Enter an upper limit of **0.0028**.
e) Enter a lower limit of **0**.
f) Number of decimal places = **4**
g) ✔

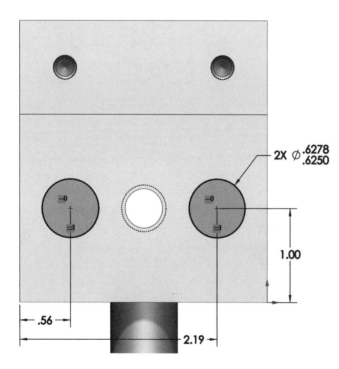
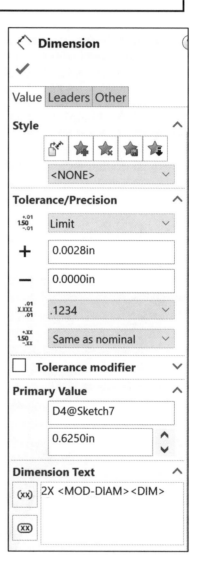

2) **IMPORTANT!! Save** and **keep** this part. It will be used in a future tutorial.

11.6) VISE - SCREW TUTORIAL

11.6.1) Prerequisites

Before starting this tutorial, you should have completed the following tutorials.

- Stationary Jaw Tutorial

You should also have the following knowledge.

- Familiarity with threads and fasteners
- Familiarity with tolerancing

11.6.2) What you will learn

The objective of this tutorial is to introduce you to realistic part models that have toleranced dimensions. You will be modeling the *Screw* from the *Vise* assembly shown in Figure 11.5-1. Specifically, you will be learning the following commands and concepts.

Sketching

- Applying tolerances

Features

- Cosmetic Thread

11.6.3) Setting up the part

1) Start a **new part** and save the part as **SCREW.SLDPRT** (**File – Save**). Remember to save often throughout this project.

2) Set the drafting standard to **ANSI** and set the text to **upper case** for notes, tables, and dimensions. (*Options* ⚙ *– Document Properties – Drafting Standard*)

3) Set the units to **IPS** (i.e. inch, pound, second) and set the **Decimals = .12**. Also, select the rounding option, **Round half to even**. (*Options* ⚙ *– Document Properties – Units*)

11.6.4) Modeling the part

1) We will be modeling the *Screw* shown in the figure on the next page. It is made from **AISI 1020 Steel, Cold Rolled**. Start by **Revolving** the following sketch. Note the location of the origin.

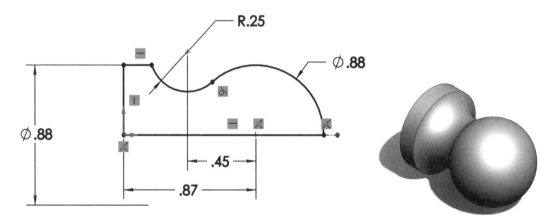

2) **Sketch** on the **Front plane** and use an **Extruded Cut** to create a hole in the head of the *Screw*.

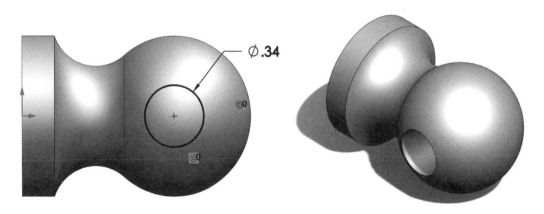

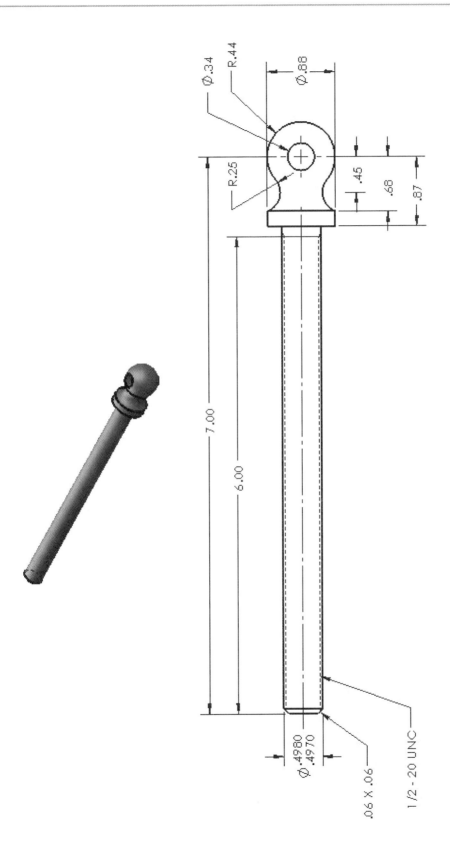

3) **Extrude** the shaft using a basic size of **0.50** inches.

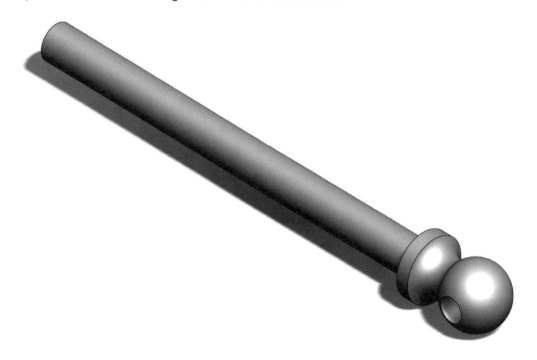

11.6.5) Applying threads

1) Apply threads to the shaft using the **Cosmetic Thread** Cosmetic Thread... command.

2) Apply a **0.06 x 0.06 Chamfer** to the end of the shaft.

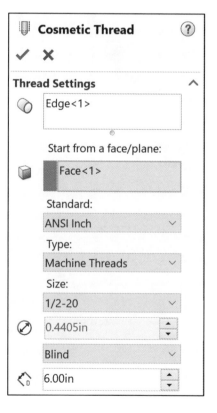

11.6.6) Applying Tolerances

1) **Edit** the **sketch** associated with the shaft **Extrude** and apply the following tolerance to the diameter dimension.

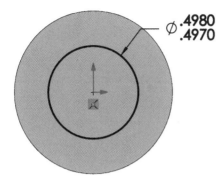 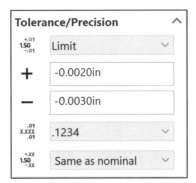

2) **IMPORTANT!! Save** and **keep** this part. It will be used in a future tutorial.

<u>NOTES:</u>

TOLERANCING AND THREADS IN SOLIDWORKS® PROBLEMS

P11-1) Model the following Cast Iron *Base*. Apply the appropriate tolerances. If you have not been taught how to calculate tolerances, ask your instructor for the appropriate values.

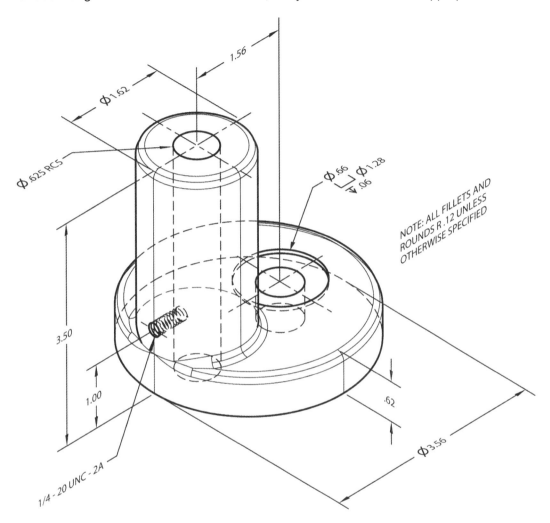

P11-2) Model the following 1045 Steel *Screw*. Apply the appropriate tolerances. If you have not been taught how to calculate tolerances, ask your instructor for the appropriate values.

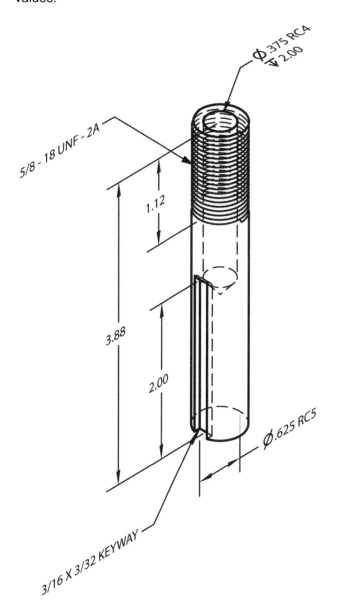

P11-3) Model the following 1045 Steel *V-Anvil*. Apply the appropriate tolerances. If you have not been taught how to calculate tolerances, ask your instructor for the appropriate values.

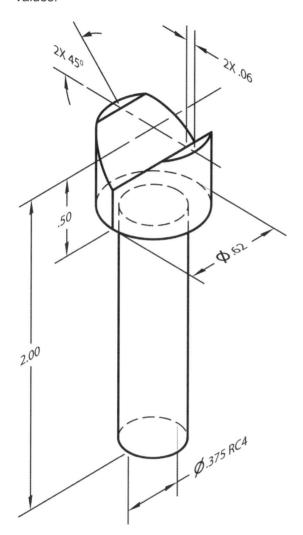

2X 45°

2X .06

.50

Ø.62

2.00

Ø.375 RC4

P11-4) Model the following 1045 Steel *Knurled Nut*. Apply the appropriate tolerances. If you have not been taught how to calculate tolerances, ask your instructor for the appropriate values.

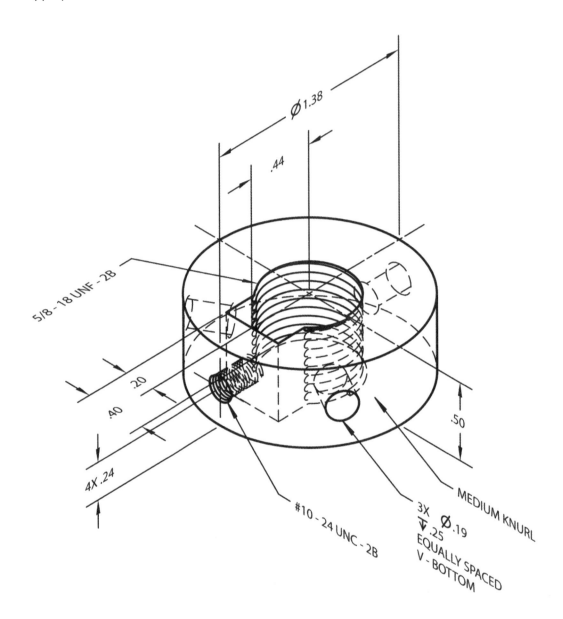

CHAPTER 12

PARAMETRIC MODELING IN SOLIDWORKS®

CHAPTER OUTLINE

12.1) PARAMETRIC MODELING 2

12.2) EQUATIONS 2

    12.2.1) Naming Dimensions 2

    12.2.2) Equations, Global Variables, and Dimensions 3

    12.2.3) Adding Dimension Equations 5

    12.2.4) Feature suppression 6

    12.2.5) IF Statement 6

12.3) VISE - SPACER TUTORIAL 7

    12.3.1) Prerequisites 7

    12.3.2) What you will learn 7

    12.3.3) Setting up the part 9

    12.3.4) Design intent 9

    12.3.5) Applying equations 11

    12.3.6) Feature suppression 14

12.4) VISE - JAW INSERT 15

    12.4.1) Prerequisites 15

    12.4.2) What you will learn 15

    12.4.3) Setting up the part 16

    12.4.4) Design intent 16

    12.4.5) Applying Equations 18

PARAMETRIC MODELING IN SOLIDWORKS® PROBLEMS 23

CHAPTER SUMMARY

In this chapter, you will learn how to create part models that have built in design intent. Parametric modeling will be used to create equations and rules that govern the parts' behavior. By the end of this chapter, you will be able to create a part that can be modified and still follow the rules that you have designed into it.

12.1) PARAMETRIC MODELING

Parametric Modeling is a modeling approach that captures design intent using features and constraints. This allows the part size to be changed but has the family of products retain the intent behind their design. **Direct modeling**, on the other hand, is a one-time-use model. This means that you model the design just for your current needs without worrying about the design intent. This is like a clay model. The clay model can't be changed after the fact. Parametric modeling supports designs that will need to be modified or integrated on a regular basis. But this type of modeling takes more planning. The design intent needs to be built in initially. When you model a part with built-in design intent, you define how the part will change when parameters are modified. In general, you want to design a part for change and flexibility. This can be accomplished using parametric modeling. Parametric modeling allows you to relate two or more feature dimensions together so that when one changes, they all change according to the design intent.

Parametric Modeling is, in the context of this book, the creation of a SOLIDWORKS® part using parameters, equations, and rules. For example, one dimension on the part could be linked to another dimension on the part through an equation (e.g., diameter1 = 2*diameter2). It could also mean that a feature appears or disappears when another dimension reaches a specified value, e.g., if (Hole diameter < 0.25, then suppress, else unsuppress).

12.2) EQUATIONS

12.2.1) Naming Dimensions

A big part of parametric modeling is linking dimensions through equations. To make that easier, we give each dimension a name that has meaning. After a part has been modeled, show the feature dimensions, view the dimension names, and then name the dimensions. The steps to accomplish these are given below.

Showing feature dimensions

To view the sketch and feature dimensions on the model, do the following.

1) Right-click on **Annotations** in the *Feature Tree*.
2) Select **Show Feature Dimensions**.

Viewing dimension names

1) From the pull-down menu, select **View – Hide/Show - Dimension Names**
or

1) In the *Heads-up View* toolbar, select the arrow next to ⌖▾ and select ▣ .

Naming dimensions

1) Click on the dimension to be named. If you click on the dimension value, you will have to double-click.
2) In the *Dimension* window, enter the dimension (see Figure 12.2-1). Note that the @Sketch will be added automatically. This is the sketch where the dimension is attached.

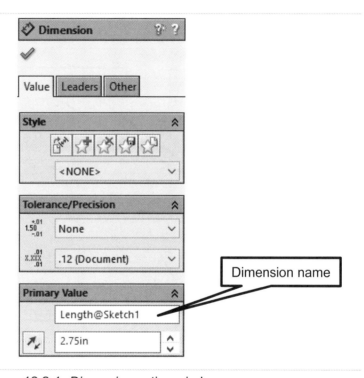

Figure 12.2-1: *Dimension* option window

12.2.2) Equations, Global Variables, and Dimensions

Equations allow you to relate dimensions to global variables or other dimensions using mathematical functions. This can be done in both parts and assemblies. The easiest way to apply equations is through the ***Equations, Global Variables, and Dimensions*** window. To access this window, use the following steps.

Equations, Global Variables, and Dimensions window

1) Select **Tools – Equations**
2) In the _Equations, Global Variables, and Dimensions_ window, you can add relationships (see Figure 12.2-2)
    a) <u>Global Variables</u> = A number assigned to a variable that, in general, will not change.   Global variables may be used to tie dimension together using equations.
    b) <u>Features</u> = Features (such as Extrudes, Holes, etc.) may be suppressed.
    c) <u>Dimensions</u> = Dimensions may be defined using equations.
or
1) From the pull-down menu, select **View – Toolbars – Tools**.
2) Select Σ.

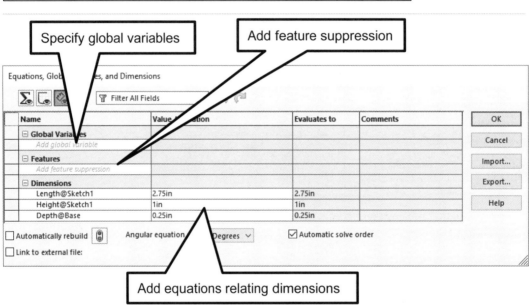

Figure 12.2-2: _Equations, Global Variables, and Dimensions_ window

12.2.3) Adding Dimension Equations

Dimension equations may be added in the *Equations, Global Variables, and Dimensions* window or in the *Dimension* options window. Use the following steps to add equations in the *Equations, Global Variables, and Dimensions* window.

Adding dimension equations

1) The *Name* column under the *Dimensions* field should show all the dimensions used to create the part. If the dimensions don't appear, click on the **Dimension View** icon  at the top.
2) Click in the cell next to the dimension you wish to define (under the *Value/Equation* column).
3) Start with an "=". This will prompt selections that allow you to enter *Functions*, *File Properties* or *Measure…*

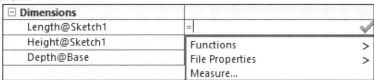

4) Follow the "=" with the desired equation. SOLIDWORKS® generally follows the equation rules and syntax of EXCEL®. Variables/Dimension names need to be enclosed in double quotes and will turn blue if acceptable. You may also just click on the dimension attached to the part and the name will automatically be added. A green check next to the equation will appear if the equation is valid.
5) The calculated value of the equation will be shown in the cell to the right of the equation under the *Evaluates to* column.

⊟ **Dimensions**		
Length@Sketch1	2.75in	2.75in
Height@Sketch1	=(4/11)*"Length@Sketch1"	✔ 1in

6) If an error occurs, that tells you that the equations are out of order. Activate the **Automatic solve order** checkbox.

⚠ Hole separation 1/2@Sketch2	= .5 * "Hole separation@Sketch2"	0.875in
⚠ Hole separation@Sketch2	= "Length@Sketch1" - 2 * "Hole DIA@Sketch2"	1.75in

☐ Automatically rebuild   Angular equation units: Degrees ∨   ☐ Automatic solve order

7) After applying an equation, the dimension should have a Σ symbol next to it.

Σ 1.00 (Height)

12.2.4) Feature suppression

Features may be suppressed or unsuppressed based on rules that are prescribed. Use the following steps to add feature suppression.

Feature suppression

1) Enter the *Equations, Global Variables, and Dimensions* window (**Tools – Equations**).
2) Click on **Add feature suppression** under *Features*.

3) Select the feature in the *Feature Design Tree* or enter or click on the feature name that is shown in the *Feature Design Tree*.
4) Click in the cell to the right of the feature (under the *Value/Equations* column). A list of options will appear. To simply suppress a feature, select **Global Variables** and then **suppress**. However, the **suppress** command is usually applied within an if statement.

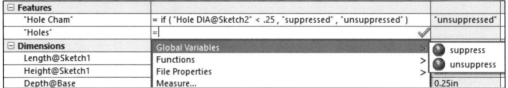

5) Functions may be used to indicate conditions for suppression.

12.2.5) IF Statement

An IF statement may be used to change the value of a dimension or to suppress a feature based on the change of another dimension. The syntax for the **IF** statement is shown in the following two examples. Mathematical symbols such as =, <, > may be used. The equation shown below states that the *Hole DIA@Sketch2* is 0.5 units if *Length@Sketch1* is greater than 2 units and 0.25 units if *Length@Sketch1* is less than or equal to 2.

Hole DIA@Sketch2	=if("Length@Sketch1" > 2, 0.5, 0.25)

The equation shown below states that if the *Hole DIA* becomes less than .25 inches, suppress the *Hole Cham*. If the *Hole DIA* becomes equal to or greater than .25, the feature is unsuppressed.

Hole Cham	=if("Hole DIA@Sketch2" < .25 in, "suppressed", "unsuppressed")

12.3) VISE - SPACER TUTORIAL

12.3.1) Prerequisites

Before starting this tutorial, you should have completed the following tutorials.

- Chapter 1 - Connecting Rod Tutorial
- Chapter 3 - Flanged Coupling Tutorial

12.3.2) What you will learn

The objective of this tutorial is to introduce you to part models that have built-in design intent using parametric modeling. You will be modeling the *Spacer* from the *Vise* assembly shown in Figure 12.3-1. Specifically, you will be learning the following commands and concepts.

Sketching

- Dimension names
- Equations

Features

- Dimension names
- Viewing feature dimensions
- Equations
- Feature suppression
- If statement

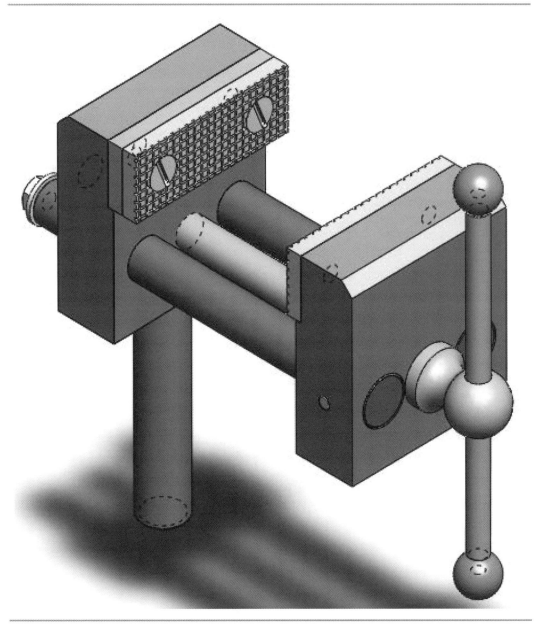

Figure 12.3-1: Vise assembly

12.3.3) Setting up the part

1) Start a **New Part**. We will be modeling the following **SPACER** which is made of **AISI 1020 Steel, Cold rolled**.

2) Set the drafting standard to **ANSI** and set the text to **upper case** for notes, tables, and dimensions. (*Options* 🔘 *– Document Properties – Drafting Standard*)

3) Set the units to **IPS** (i.e. inch, pound, second) and set the **Decimals = .123**. Also, select the rounding option, **Round half to even**. (*Options* 🔘 *– Document Properties – Units*)

12.3.4) Design intent

1) We will be building in the following design intent.
    a) The height and depth are related to the length. Which means that if the length is changed, the height and depth will change.
    b) No matter what size the part is, the holes will remain horizontally centered.
    c) The location of the holes will depend on the height of the part.
    d) The size of the holes will depend on the length of the part.

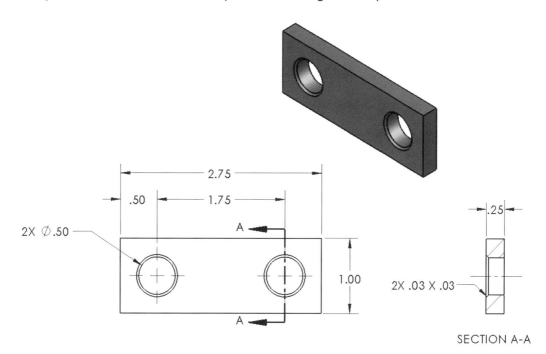

2) Start by drawing the following rectangle with the dimensions shown and named.

   a) **Sketch** and **Dimension** the **Rectangle**  shown on the **Front plane**.

   b) In the *Heads-up View* toolbar, select the arrow next to [👁▾] and select [D1].

   c) Click on a dimension to be named. If you click on the dimension value, you will need to double-click. In the *Dimension* window, enter the dimension. Note that the @Sketch will be added automatically after the dimension name.

Primary Value	^
Length@Sketch1	
↗ 2.750in	⌄

   ➢ See section 12.2.1 to learn about ***Showing, Naming, and Viewing Dimensions***.

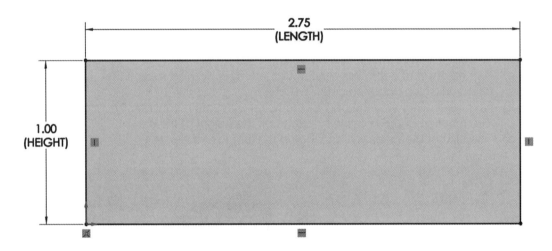

3) **Extrude** [Extruded Boss/Base] the sketch to **0.25 inches**. Name the *Extrude*, **Base**. (Remember, to name features, double-click slowly on the feature name in the *Feature Design Tree*.)

12.3.5) Applying equations

1) Right click on **Annotations** in the *Feature Design Tree* and select **Show Feature Dimensions**. If the dimensions don't appear, click on the *Base* extrude.

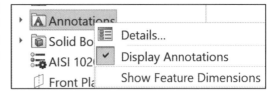

2) Click on the 0.25 dimension and name it **Depth**.

3) Apply the following equations (**HEIGHT = (4/11)*LENGTH, DEPTH = (1/11)*LENGTH**).
   a) From the pull-down menu, select **View – Toolbars – Tools**.
   b) Select $\boxed{\Sigma}$ from the *Tools* toolbar.
   c) In the *Equations, Global Variables, and Dimensions* window, enter the following equations next to the respective dimension names. Start the equation with an "=" sign. Note that the dimension name is in double quotes. An easy way to insert the dimension name is to just click on it within the modeling area.
   d) **OK**
   e) Note that the *HEIGHT* and *DEPTH* dimensions now have Σ symbols next to them.

> ➤ See section 12.2.2 and 12.2.3 to learn about **Adding Dimension Equations**.

− Dimensions		
LENGTH@Sketch1	2.75in	2.75in
HEIGHT@Sketch1	= (4 / 11) * "LENGTH@Sketch1"	1in
DEPTH@Base	= (1 / 11) * "LENGTH@Sketch1"	0.25in

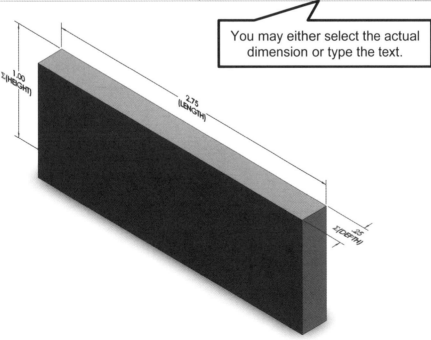

You may either select the actual dimension or type the text.

4) **Sketch** and **Dimension** the following **Centerlines**  and **Circles** on the front face of the part. Apply the following **Sketch Relations, Dimension names,** and **Equations.** Then, **Extrude Cut Through All** and name the Extrude Cut **Holes.**
   a) The **Centerlines** start and end on the **Midpoint** of the sides.
   b) The **Circle centers** are aligned on the horizontal **Centerline.**
   c) The two **Circles** are **Equal.**

5) **Name** the **dimensions** as shown in the figure.

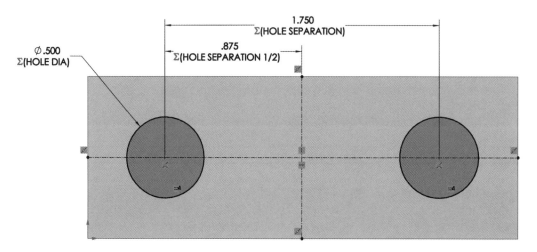

6) **Add** the **equations** shown below.

− Dimensions		
LENGTH@Sketch1	2.75in	2.75in
HEIGHT@Sketch1	= (4 / 11) * "LENGTH@Sketch1"	1in
DEPTH@Base	= (1 / 11) * "LENGTH@Sketch1"	0.25in
HOLE DIA@Sketch2	= .5 * "HEIGHT@Sketch1"	0.5in
HOLE SEPARATION 1/2@Sketch2	= .5 * "HOLE SEPARATION@Sketch2"	0.875in
HOLE SEPARATION@Sketch2	= "LENGTH@Sketch1" - 2 * "HOLE DIA@Sketch2"	1.75in

7) **Exit sketch.**

8) Change **Length** to **4 inches**. The dimensions should change automatically in accordance with the applied equations. You may have to **update** 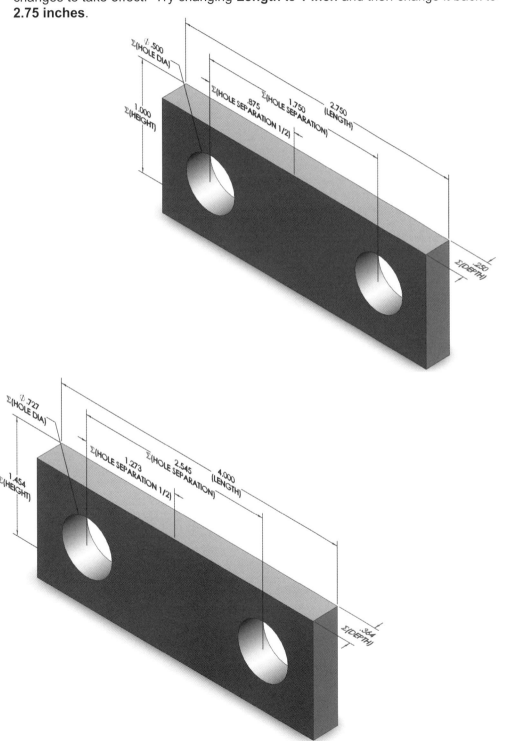 for the changes to take effect. Try changing **Length** to **1 inch** and then change it back to **2.75 inches**.

12.3.6) Feature suppression

1) Apply a **0.03 X 0.03 Chamfer** 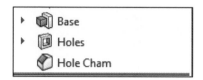 to the front face of the holes and name the dimensions as shown.

2) Name the *Design Tree* features as shown.

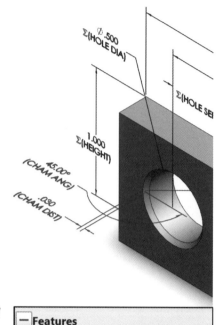

▸ 🔲 Base
▸ 🔲 Holes
　🔷 Hole Cham

3) Suppress the **Hole Cham** if the **HOLE DIA** becomes **less than 0.25 inches**.

   a) Select ⟦Σ⟧ from the *Tools* toolbar.
   b) Under *Features*, select the cell that says *Add feature suppression*.

− Features
Add feature suppression

   c) Select **Hole Cham** in the *Feature Tree*.
   d) A window will appear in the next column. Select **Functions** and then **if()**.
   e) Click inside the parentheses and then click on the **HOLE DIA** dimension. Make sure there is no space in front of the text.
   f) After the dimension name, type **<.25,**
   g) Another window will appear, select **Global** and then **suppressed**, then type a comma, and then select **Global** and then **unsuppressed**.
   h) **Ok**
   i) Change **Length** to **1 inch** and see if the chamfer is suppressed. Change **Length** back to **2.75 inch**.

− Features		
"Hole Cham"	= IIF ("HOLE DIA@Sketch2" < .25 , "suppressed" , "unsuppressed")	Unsuppressed

> ➤ See section 12.2.4 and 12.2.5 to learn about **Suppressing Features** and the **IF Statement**.

4) On your own, relate the *CHAM DIST* to the *HOLE DIA*. If the *HOLE DIA* is **less than 0.75 inch** then the *CHAM DIST* is **0.03 inch**. If the *HOLE DIA* is **greater than or equal to 0.75 inch**, then the *CHAM DIST* is **0.06 inch**. Change **Length** to **5.00** inches and then to **1.00** inch to see if the equations work. Change **Length** back to **2.75** inches.

5) **IMPORTANT!! Save** and **keep** this part. It will be used in a future tutorial.

12.4) VISE - JAW INSERT

12.4.1) Prerequisites

Before starting this tutorial, you should have completed the following tutorials.

- Vise - Spacer

It will help if you have the following knowledge.

- Familiarity with threads and fasteners

12.4.2) What you will learn

The objective of this tutorial is to introduce you to part models that have built-in design intent using parametric modeling. You will be modeling the *Jaw Insert* from the *Vise* assembly shown in Figure 12.3-1. Specifically, you will be learning the following commands and concepts.

Sketching

- Dimension names
- Equations

Features

- Dimension names
- Viewing feature dimensions
- Equations

12.4.3) Setting up the part

1) Start a **New Part.** We will be modeling the following **JAW INSERT** which is made of **Alloy steel**.

2) Set the drafting standard to **ANSI**. (*Options* ⚙ – *Document Properties – Drafting Standard*)

3) Set the units to **IPS** (i.e. inch, pound, second) and set the **Decimals = .123**. Also, select the rounding option, **Round half to even**. (*Options* ⚙ – *Document Properties – Units*)

12.4.4) Design intent

1) We will be building in the following design intent.
   a) The number of countersunk holes will depend on the length of the part with equal spacing between them.
   b) Regardless of the size of the part, the holes will remain vertically centered.
   c) The grooves will cover the entire face of the part regardless of its size.

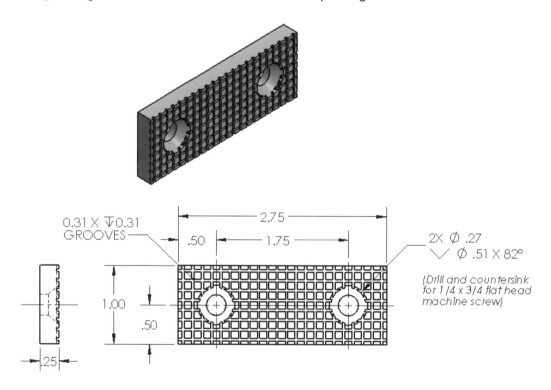

2) **Sketch** and **Dimension** the following **Rectangle** 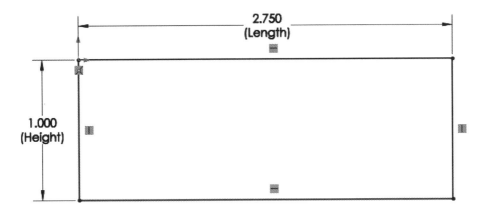 on the **Front plane**. Name the dimensions as shown. **Extrude** the sketch to a thickness of **0.25 inch**.

3) Apply the material **Alloy Steel** and save the part as **JAW INSERT**.

4) Use the Hole Wizard 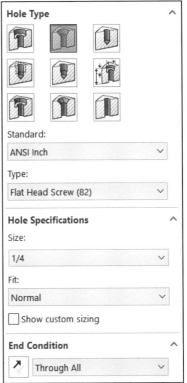 to insert a **Countersunk** normal clearance hole for a **1/4 flat head screw**. Position the hole as shown. Note that the **Centerline** is constrained to the **Midpoints** of the sides and the center of the hole is **Coincident** with the centerline.
   a) Standard = **ANSI Inch**
   b) Type: = **Flat Head Screw (82)**
   c) Size = **1/4**
   d) Fit = **Normal**
   e) End Condition = **Through All**

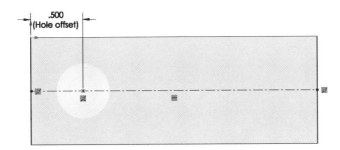

5) Create a vertical groove using a **Rectangle** ▢ Corner Rectangle on the front face of the part. Name the dimensions as shown. **Extrude Cut** the rectangle **0.031 inch**.

6) Right click on **Annotations** in the *Feature Design Tree* and select **Show Feature Dimensions**.

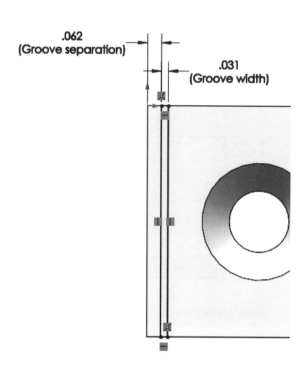

12.4.5) Applying Equations

1) Create a horizontal groove similar to the vertical groove using a **Rectangle** ▢ Corner Rectangle on the front face of the part. Name the dimensions as shown in the figure. **Extrude Cut** the rectangle **0.031 inch**.

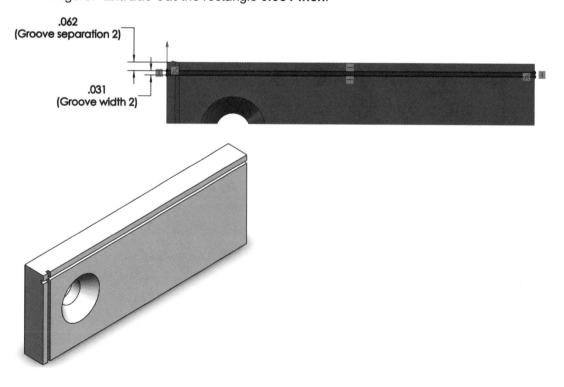

2) Name the features in the *Feature Manager Design Tree* as shown.

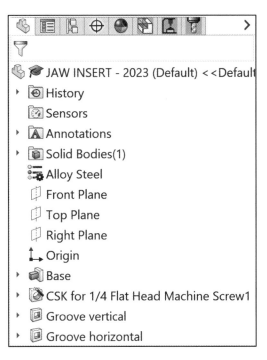

3) Add the **Equations** shown below.

Goove separation@Sketch4	0.062in	0.062in
Groove width@Sketch4	0.031in	0.031in
D1@Groove vertical	0.031in	0.031in
Groove separation 2@Sketch5	= "Goove separation@Sketch4"	0.062in
Groove width 2@Sketch5	= "Groove width@Sketch4"	0.031in
D1@Groove horizontal	= "D1@Groove vertical"	0.031in

4) **Pattern** the countersunk hole **1.75 in** apart using the following equations to control the number of holes (**=int("Length@Sketch1"/1.75 + 0.5)**). The int() function returns an integer.

5) **Pattern** *Groove vertical* using the following equations to control the number of grooves (**=int("Length@Sketch1"/0.093)**). Set the distance between the groove equal to **=3*"Groove width@Sketch4"**.

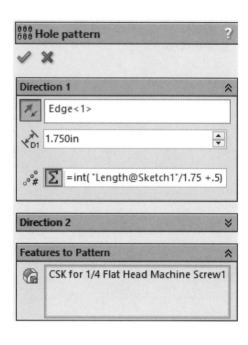

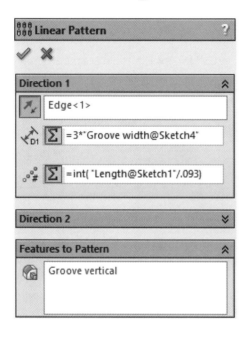

6) **Pattern** *Groove horizontal* using the following equations to control the number of grooves (**=int("Height@Sketch1"/0.093)**). Set the distance between the groove equal to **=3*"Groove width@Sketch4"**.

7) Change the **Length** to **5 inches** and the **Height** to **2 inches** to see the effect of the equations. You may have to **update** 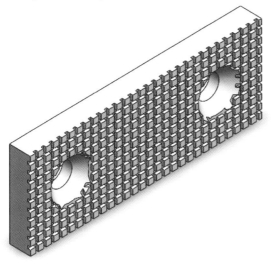 to see the effects.

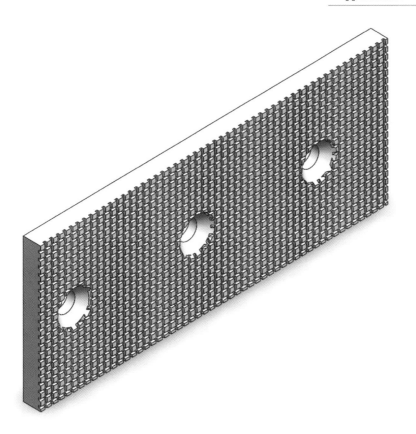

<u>NOTES:</u>

PARAMETRIC MODELING IN SOLIDWORKS® PROBLEMS

P12-1) Use SOLIDWORKS® to create the following part using the stated design intent requirements.
1) The holes always remain centered with respect to the width (50 mm).
2) The 10 mm width of the part changes uniformly all around.
3) The diameters of the 3 holes are proportional to the width of the part.
4) The thickness of the part (10 mm, all the way around the part) is proportional to the length of the part (110).
5) Challenge: The toleranced hole changes appropriately as the nominal diameter of the hole changes.

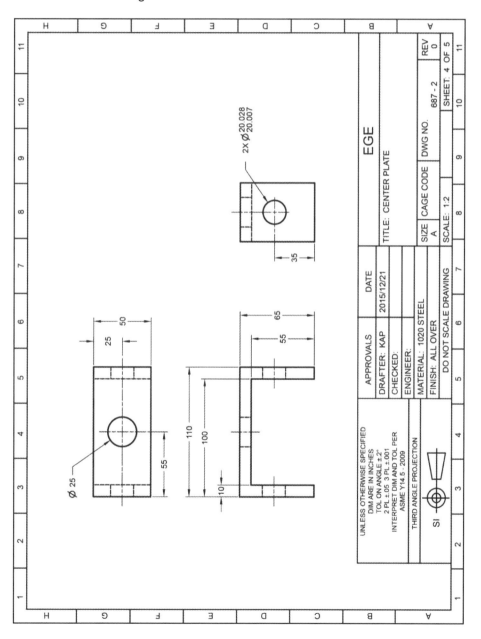

P12-2) Use SOLIDWORKS® to create the following part using the stated design intent requirements.
   1) The holes remain centered with respect to the height of the part (2.00 in).
   2) The number of holes increase as the length (4.50 in) allows. The distance between each successive hole should remain the same.

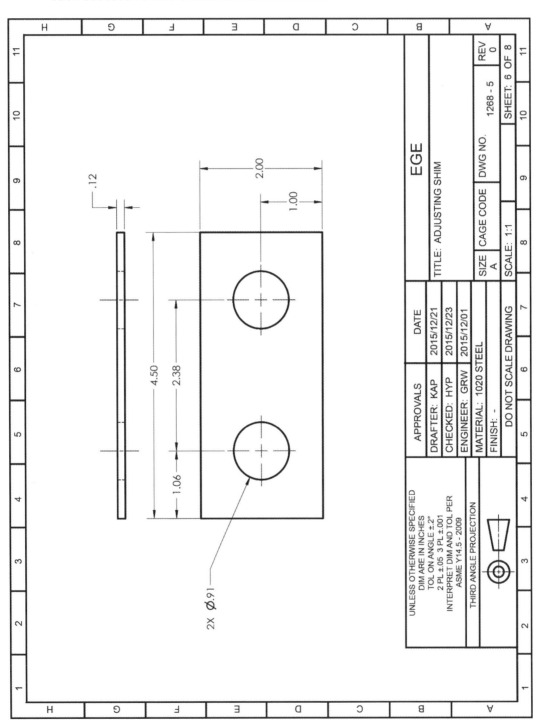

CHAPTER 13

ADVANCED ASSEMBLIES IN SOLIDWORKS®

CHAPTER OUTLINE

13.1) ADDING PARTS TO AN ASSEMBLY 2

13.2) MATES 2

    13.2.1) The Mate command 2

    13.2.2) Mechanical Mates 3

13.3) VISE ASSEMBLY TUTORIAL 5

    13.3.1) Prerequisites 5

    13.3.2) What you will learn 5

    13.3.3) Inserting components 7

    13.3.4) Standard mates 8

    13.3.5) Mechanical mates 12

    13.3.6) Toolbox components 13

    13.3.7) Advanced mates 15

13.4) GEAR ASSEMBLY TUTORIAL 18

    13.4.1) Prerequisites 18

    13.4.2) What you will learn 18

    13.4.3) Setting up the Assembly 20

    13.4.4) Adding gears 20

    13.4.5) Adding a belt 24

    13.4.6) Gear mate 25

    13.4.7) Rack Pinion mate 26

    13.4.8) Bevel Gear 28

    13.4.9) Inserting a New Part 30

ADVANCED ASSEMBLIES IN SOLIDWORKS® PROBLEMS 35

CHAPTER SUMMARY

In this chapter, you will learn how to create realistic assemblies in SOLIDWORKS®. SOLIDWORKS® has many different types of mates that can be used to assemble parts. This chapter will focus on standard, advanced, and mechanical mates. By the end of this chapter, you will be able to create complex assemblies that behave in a realistic manner.

13.1) ADDING PARTS TO AN ASSEMBLY

In the tutorials so far, we have had all the parts opened that are included in the assembly. This makes it very easy to choose the parts in the *Begin Assembly* options window. However, there are other ways of inserting or adding a part to an assembly. When you are in an assembly model, you can insert components using the following commands.

- **Insert Components:** 🗐 Insert Components   This command enables an insertion of a part that has already been created.
- **New Part:** 🗐 New Part   This allows for a creation of a new part in the context of the assembly and the use of existing geometry to design a part. When creating a new part within an assembly an **InPlace** (coincident) mate is automatically created.
- **New Assembly:** 🗐 New Assembly   A new subassembly within the assembly can be created.
- **Copy with Mates:** 🗐 Copy with Mates   This is used to copy an existing component in an assembly and include its mates.

13.2) MATES

Mates allow physical constraints between parts in an assembly to be created. For example, making surface contacts or making two shafts run along the same axis. There are three categories of mates: *Standard, Advanced, and Mechanical*.

13.2.1) The Mate command

The **Mate** 🗐 Mate   command is located in the *Assembly* tab. Figure 13.2-1 shows the *Mate* options window and the available mates. Chapter 7 and 10 covered *Standard* and *Advanced* mates and how to apply them. This chapter will focus on *Mechanical* mates. *Mechanical* mates can be tricky to apply and will need to be practiced often in order to become proficient at them.

13.2.2) Mechanical Mates

Along with the *Standard* and *Advanced* mates, there are a set of *Mechanical* mates. SOLIDWORKS® gives a variety of *Mechanical* mates to choose from. These mates are associated with certain machine elements. The available *Mechanical* mates are shown in Figure 13.2-1 and described below.

- **Cam:** This mate is a type of coincident mate. It allows a cylinder, plane, or point to a surface that is closed and does not have any discontinuities to be mated.

- **Slot:** This mate forces an axis or cylinder to move within a slot.

- **Hinge:** This mate limits the movement between two components to one rotational degree of freedom. The angular movement between the two components may also be limited.

- **Gear:** This mate forces two components to rotate relative to one another about selected axes.

- **Rack Pinion:** This mate forces one component to rotate when another translates or vice versa.

- **Screw:** This mate constrains two components to be concentric and also adds a pitch relationship. This means that the translation of one component along the axis causes rotation of the other component according to the pitch relationship or vice versa.

- **Universal Joint:** In this mate, the rotation of one component about its axis is driven by the rotation of another component about its axis. The axes usually point in different directions.

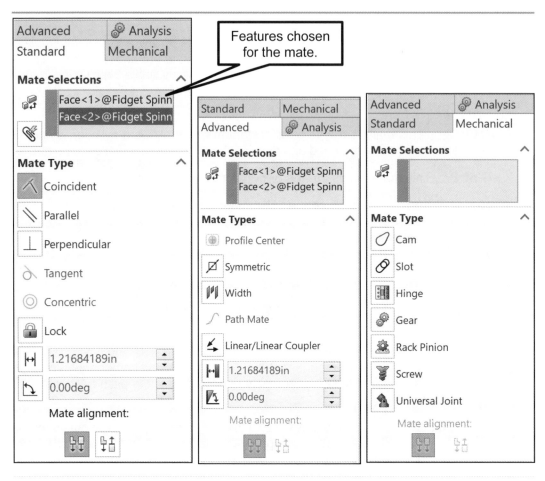

Figure 13.2-1: *Mate* Options Window

13.3) VISE ASSEMBLY TUTORIAL

13.3.1) Prerequisites

Before starting this tutorial, you should have completed the following tutorials.

- Vise – Stationary Jaw Tutorial
- Vise – Screw Tutorial
- Vise – Spacer Tutorial
- Vise – Jaw Insert Tutorial
- Flange Coupling Assembly Tutorial
- Linear bearing Assembly Tutorial

It will help if you have the following knowledge.

- A familiarity with threads and fasteners.

13.3.2) What you will learn

The objective of this tutorial is to introduce you to *Advanced* and *Mechanical Mates*. You will assemble the parts that comprise the *Vise* assembly shown in Figure 13.3-1. Specifically, you will be learning the following commands and concepts.

Assembly

- Insert Component
- Mechanical Mates – Screw mate

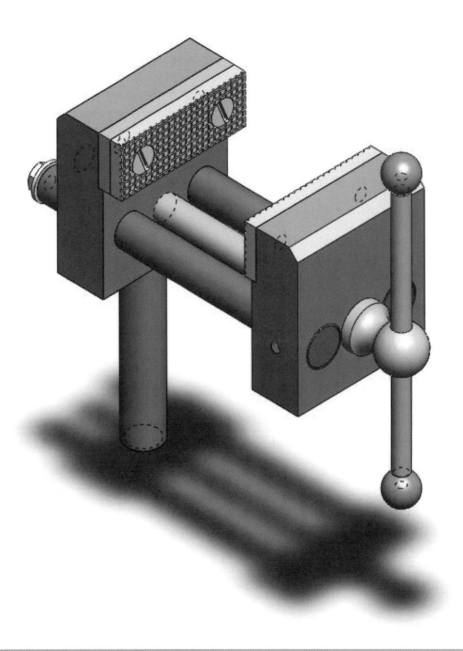

Figure 13.3-1: Vise assembly

13.3.3) Inserting components

1) **Download** the following parts and add them to the parts that have already been created in Chapter 11 and 12 (i.e., **STATIONARY JAW, SCREW, SPACER, JAW INSERT**).
   - **GUIDE BAR.SLDPRT**
   - **HANDLE.SLDPRT**
   - **MOVABLE JAW.SLDPRT**
   - **REMOVABLE BALL.SLDPRT**

2) Start a **New Assembly** .

3) **Cancel** the *Open* window and the *Begin Assembly* window.

4) Set the units to **IPS**.

5) **Save** the assembly as **VISE.SLDASM** (**File – Save**).   Remember to save often throughout this tutorial.

6) In the *Assembly* tab select **Insert Components**  and insert the **STATIONARY JAW** into the assembly. Because the *STATIONARY JAW* is the first component, it will be fixed.

   - See section 13.1 to learn about ***Inserting Components***.

7) Use **Insert Components**  to insert the following parts.
   - **MOVABLE JAW**
   - **STATIONARY JAW**
   - **SCREW**
   - **GUIDE BAR**
   - **HANDLE**
   - **REMOVABLE BALL**
   - **SPACER**
   - **JAW INSERT**

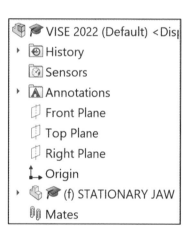

8) Make a **copy** of the following parts (**Ctrl + click + drag**).
   - **GUIDE BAR**
   - **REMOVABLE BALL**
   - **JAW INSERT**

13.3.4) Standard mates

1) If the *STATIONARY JAW* is not in the position shown do the following.
   a) **Float** the *STATIONARY JAW* and apply the mates in the next step (right click on the part in the *Feature Design Tree* and select *Float*).

   b) Apply the following **Standard Mate** [Mate] to the **Stationary Jaw**. Make the face of the *Stationary Jaw* that contains the holes **Coincident** with the **Right Plane** of the assembly. Flip the alignment, if need be, so that the stepped face can be seen. Then make the perpendicular face **Coincident** to the **Front Plane** of the assembly.
   c) **Fix** the *Stationary Jaw* (right click on the part in the *Feature Design Tree* and select either *Fix*).

2) Apply the following **Standard Mates** 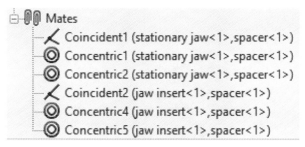 between the **STATIONARY JAW**, **SPACER**, and **JAW INSERT** to achieve the following. The *SPACER* and *JAW INSERT* should align with the holes in the *STATIONARY JAW*.
   - **Coincident** between faces (Stationary jaw, Spacer)
   - **Concentric** between both holes (Stationary jaw, Spacer)
   - **Coincident** between faces (Spacer, Jaw insert)
   - **Concentric** between both holes (Spacer, Jaw insert)

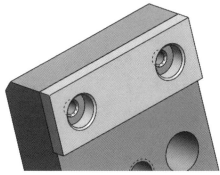

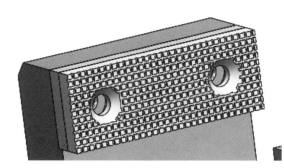

3) Apply the following **Standard Mates** to achieve similar results between the **MOVABLE JAW** and the other **JAW INSERT**. You may have to use **Mate Alignment** to get the part flipped in the correct direction.

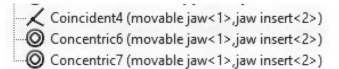

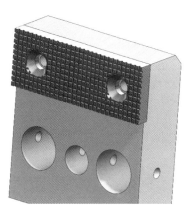

4) Apply the following **Standard Mates** Mate between the **MOVABLE JAW**, and the **GUIDE BARS** so that the *Guide bars* fit into the holes and the holes for the *Pin* align.

◎ Concentric7 (movable jaw<1>,guide bar<2>)

◎ Concentric8 (guide bar<1>,movable jaw<1>)

◎ Concentric9 (guide bar<1>,movable jaw<1>)

◎ Concentric10 (movable jaw<1>,guide bar<2>)

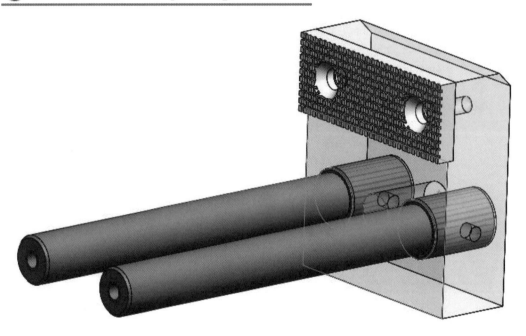

5) Apply the following **Standard Mates** [Mate] between the **STATIONARY JAW** and the **GUIDE BARS** so that the *GUIDE BARS* fit into the holes.

◎ Concentric12 (stationary jaw<1>,guide bar<2>)

◎ Concentric13 (guide bar<1>,stationary jaw<1>)

- **Having Issues?** If the second Concentric mate doesn't want to apply this could be because of the tolerances applied to the holes. You can pick two faces to be **Parallel**. One on the *Stationary Jaw* and one on the *Movable Jaw* as an alternative mate.

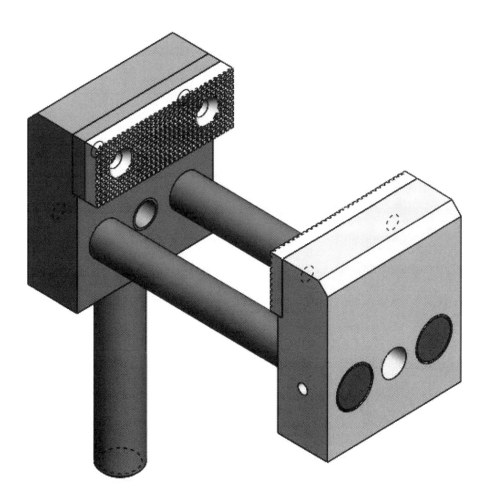

6) Apply the following **Standard Mates** 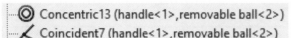 between the **HANDLE** and the **REMOVABLE BALLS**. Note that the **Coincident** mates are between the rim of the hole on the *REMOVABLE BALL* and the stepped surface on the *HANDLE*. You may have to use **Mate Alignment** to get one of the balls to be flipped in the correct direction.

◎ Concentric13 (handle<1>,removable ball<2>)
✗ Coincident7 (handle<1>,removable ball<2>)
◎ Concentric14 (handle<1>,removable ball<1>)
✗ Coincident8 (handle<1>,removable ball<1>)

13.3.5) Mechanical mates

1) Apply a **Screw Mate** between the **SCREW** and **STATIONARY JAW**.

   a. Select the **Mate** [Mate] command.
   b. Click on the *Mechanical Mates* tab.
   c. Select the **shaft** of the **SCREW** and the inner surface of the center threaded **hole** of the **STATIONARY JAW**.
   d. Select the **Screw** [Screw] mate. An arrow should appear on the *SCREW* indicating the direction of rotation. If this arrow is correct for a right-handed thread, leave it. If not, *Reverse* the direction.
   e. Select the **Revolutions/in** radio button. Since the threads are 1/2 – 13, the Screw will advance 1 inch per every **13** revolutions.

   ⌨ Screw
   ● Revolutions/in
   ○ Distance/revolution
   13
   ☑ Reverse

   f. ✓.
   g. ✓. Grab on the head of the *Screw* and rotate it. It should advance.

• See section 13.2 for information on *Mechanical Mates*.

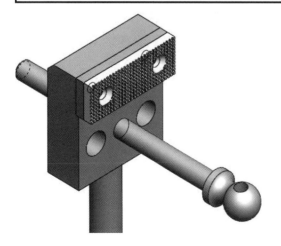

13.3.6) Toolbox components

1) Use the **Toolbox** to insert the following components.
   a) 4X **1/4 – 20 UNC Countersunk Flat Head Screw** that is 1.00 inch long with a thread length of 0.75 inch.
   b) 2X **1/4 – 20 UNC Hex Head Bolt** that is 0.875 inch long with a thread length of 0.75 inch.
   c) 2X **3/8 - Type A Plain Washer**. This will produce a 0.406 ID and 0.812 OD (Preferred Narrow).
   d) 2X **3/16 Spring Pin Slotted** that is 0.875 inch long.

- See Chapter 7 for information on **Toolbox Components**.

2) Apply the appropriate **Mates** to the toolbox components to place them into position. Use the assembly drawing shown on the next page as a guide for the placement of the toolbox component. We will mate the *HANDLE* and get the assembly in the appropriate position in the next steps.

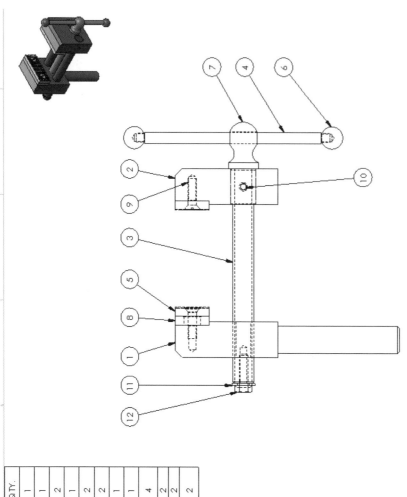

ITEM NO.	PART NUMBER	DESCRIPTION	QTY.
1	STATIONARY JAW		1
2	MOVABLE JAW		1
3	GUIDE BAR		2
4	HANDLE		1
5	JAW INSERT		2
6	REMOVABLE BALL		2
7	SCREW		1
8	SPACER		1
9	1/4 - 20 UNC FLAT HEAD SCREW		4
10	3/16 PIN		2
11	FLAT WASHER		2
12	1/4 - 20 UNC HEX HEAD BOLT		2

13.3.7) Advanced mates

1) Apply the following **Standard Mates** 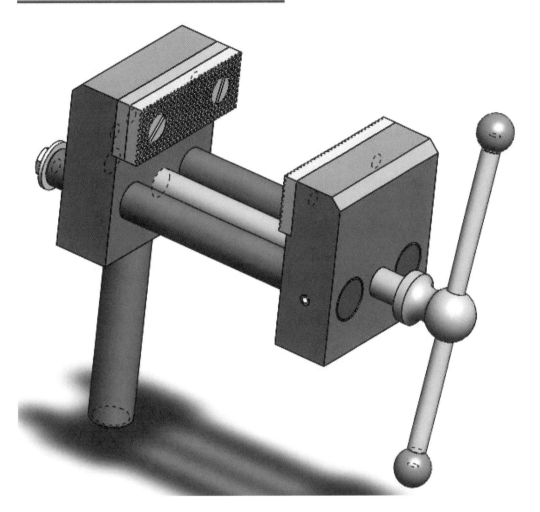 between the **HANDLE** and the **SCREW** so that the *HANDLE* fits through the hole in the *Screw*.

⊚ Concentric15 (screw<1>,handle<1>)

2) Pull on the *MOVABLE JAW, HANDLE,* and *SCREW* and see how they move in unrealistic ways. Parts will pass right though each other. This should not happen in a real assembly. We will fix these problems with *Advanced Mates*.

3) Apply a **Distance Limit Mate** to the **Washers** so that they don't go through the **Stationary Jaw**.

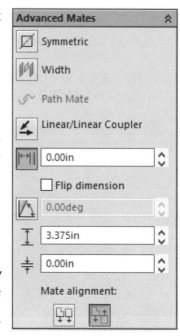

a) Select the **Mate** command.
b) Click on the *Advanced Mates* tab.

c) Select the **Distance Limit** mate.
d) Select the inner face of one of the *Washers* and the back face of the **STATIONARY JAW**, the two surfaces that would normally touch in the real assembly.
e) Set the default distance to **0 inch**.
f) Set the minimum distance to **0 inch**.
g) Set the maximum distance to **3.375 inches**.

h) ✓
i) After applying the mate, pull on the *MOVABLE JAW* to see if the mate works properly. The *Washers* should not go through the *STATIONARY JAW*.
j) The **Flip dimension** may need to be activated or deactivated if it is moving in the wrong direction.

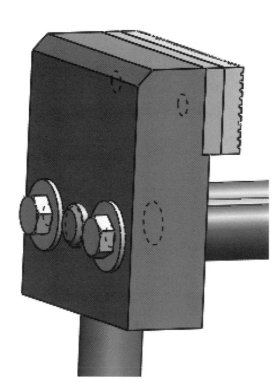

4) Apply another **Distance Limit Mate** 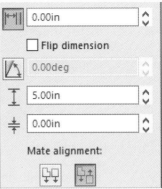 to the **SCREW** so that it does not go through the **MOVABLE JAW**. Use a default distance of **0** inches, a minimum distance of **0** inches, and a maximum distance of **5** inches.

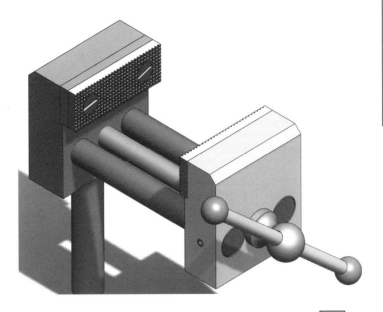

5) On your own, apply a **Distance Limit Mate** 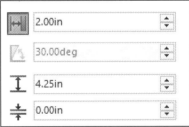 between the **HANDLE** and **SCREW**. (Select the *BALL* and the round end of the *SCREW*.)

6) See how the assembly works. The *HANDLE* may act a little wonky.

7) **Save**.

13.4) GEAR ASSEMBLY TUTORIAL

13.4.1) Prerequisites

Before starting this tutorial, you should have completed the following tutorials.

- Flange Coupling Assembly Tutorial
- Linear bearing Assembly Tutorial

It will help if you have the following knowledge.

- A familiarity with gears.

13.4.2) What you will learn

The objective of this tutorial is to introduce you to *Mechanical Mates*. In this tutorial, you will create the *Gear system* shown in Figure 13.4-1. Specifically, you will be learning the following commands and concepts.

Toolbox

- Spur Gear
- Rack Spur
- Bevel Gear

Assembly

- Gear mate
- Rack Pinion mate
- Belt/Chain
- New Part

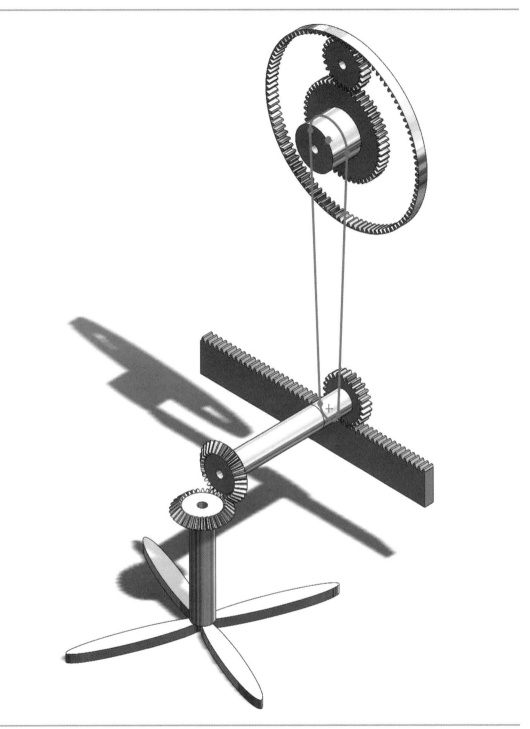

Figure 13.4-1: Gear assembly

13.4.3) Setting up the Assembly

1) Start a **New Assembly** .

2) **Cancel** the *Open* window and the *Begin Assembly* window.

3) Set the units to **MMGS**.

4) **Save** the assembly as **GEAR.SLDASM** (**File – Save**). Remember to save often throughout this tutorial.

5) On the **Front plane, Sketch** and **Dimension** the following profile. Note that the two upper circles are **Tangent**. Make sure that the sketch is completely constrained (i.e. black). When done, **Exit Sketch**.

13.4.4) Adding gears

1) Enter the **Toolbox Library** . If *Toolbox* is not added, add it now.

2) Open the *Toolbox* library and click on **Gears** under **ANSI Metric – Power Transmission**.

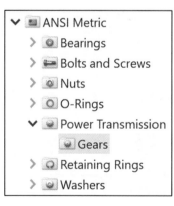

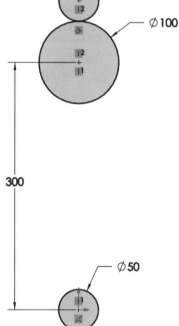

3) Insert one **Spur Gear** with the following properties.

• See Chapter 7 for information on ***Toolbox Components***.

4) **Float** the gear and then make the following **Mates** 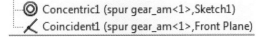.
   a) **Concentric** constraint between the **Gear** and the **large circle**.
   b) **Coincident** constraint between the **gear's front face** and the **Front plane**.

 Concentric1 (spur gear_am<1>,Sketch1)
 Coincident1 (spur gear_am<1>,Front Plane)

5) Insert two **Spur Gears** with the following properties.

6) Edit the properties of one of the small gears (right click on the part in the *Feature Manager Design Tree* – **Edit toolbox component**). Add a **hub** with a diameter of **25** and an overall length of **50**.

7) Make the following **Mates** .
   a) **Concentric** constraint between the **Gears** and the **small circles**.
   b) **Coincident** constraint between the **gear's front face** and the **Front plane**.

◎ Concentric2 (spur gear_am<4>,Sketch1)
✕ Coincident2 (spur gear_am<4>,Front Plane)
◎ Concentric3 (spur gear_am<3>,Sketch1)
✕ Coincident3 (spur gear_am<3>,Front Plane)

8) Insert an **Internal Spur Gear** with the following properties.

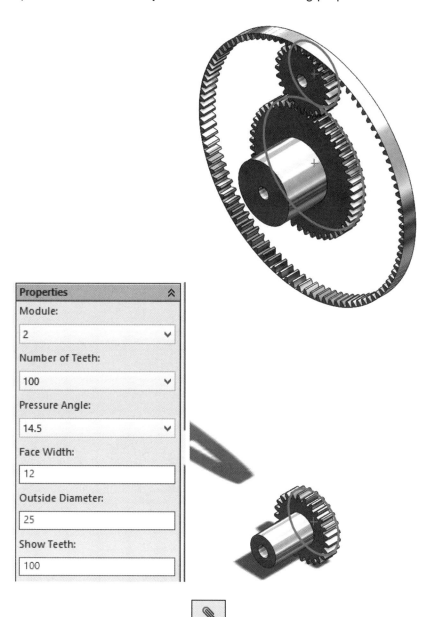

Properties	≫
Module:	
2	∨
Number of Teeth:	
100	∨
Pressure Angle:	
14.5	∨
Face Width:	
12	
Outside Diameter:	
25	
Show Teeth:	
100	

9) Make the following **Mates** .
   - **Concentric** constraint between the **Internal Gear** and the **large circle**.
   - **Coincident** constraint between the **gear's front face** and the **Front plane**.

13.4.5) Adding a belt

1) Insert a **Belt/Chain** [⚙️ Belt/Chain].
   a) Expand the **Assembly Feature** command located in the *Assembly* tab.
   b) Select **Belt/Chain**.
   c) Select the two outer circumferences of the hubs.
   d) ✔️

Assemb...	Referen...	New Mot Study

Hole Series
Hole Wizard
Simple Hole
Extruded Cut
Revolved Cut
Swept Cut
Fillet
Chamfer
Weld Bead
Belt/Chain

2) Rotate the top gear and notice that the bottom gear moves.

13.4.6) Gear mate

3) Use a **Gear mate** with a **2:1** gear ratio to mate the top large gear with the top smaller gear.

a) Align the gear teeth as shown in the figure.

b) Select **Mate** .

c) Select the *Mechanical* tab.

d) Select the **Gear** mate.

e) <u>Mate Selections:</u> Select the **inside** of the **hole** of the small gear and then the **inside** of the **hole** of the large gear.

f) Enter a gear ratio of **1:2**.

g) ✓

h) Test the mate to see if it works. The mate may have to be *Reversed* or the order of the *Ratio* may need to be changed.

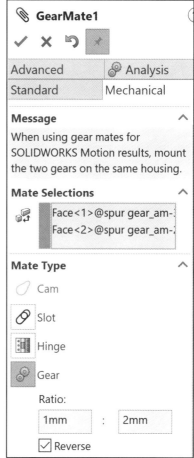

4) Use a similar method to apply a **4:1 Gear mate** between the **Internal Spur Gear** and **Small Gear**. The one difference is that for the *Internal Spur Gear*, you will choose the outer circumference as the *Mate Selection*.

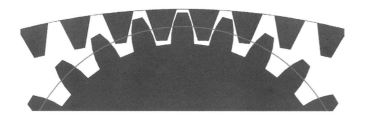

13.4.7) Rack Pinion mate

1) From the **Toolbox**, insert a **Rack Spur** with the following properties.

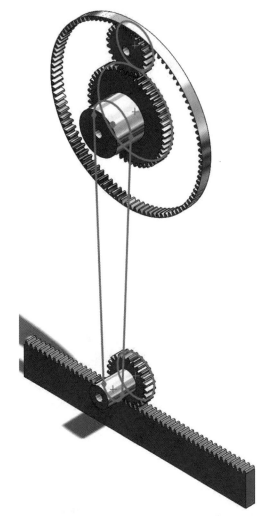

Properties ⌃

Module:

2 ⌄

Pressure Angle:

14.5 ⌄

Face Width:

12

Pitch Height:

40

Length:

300

Show Teeth:

All ⌄

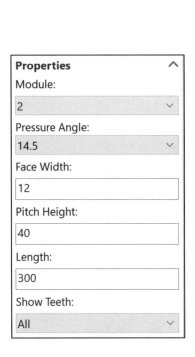

2) Apply the following **Standard Mates** .
   a) **Coincident** constraint between the long face of the **Rack** and the **Front plane**.
   b) **Parallel** mate between the bottom surface of the **Rack** and the **Top plane**.
   c) **Distance** mate of **23 mm** between the **Top Plane** and the **top** of one of the **Rack teeth**.

3) Apply a **Rack Pinion** mate between the **Rack** and the **bottom gear**.
   a) Align the *Rack* and bottom gear teeth.

   b) Select **Mate** .
   c) Select the *Mechanical* tab.

   d) Select the **Rack Pinion** mate.
   e) Rack: Select the **bottom long edge** of the *Rack*.
   f) Pinion/Gear: Select the **inside hole** surface of the *Small Gear*.
   g) Select the **Pinion pitch diameter** radio button and enter **50 mm**.

   h) ✓
   i) Test the mate to see if it works. The mate may have to be *Reversed*.

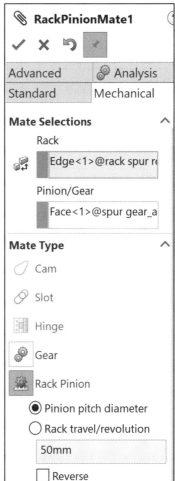

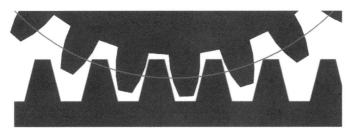

4) **Save**

13.4.8) Bevel Gear

1) From the **Toolbox**, insert **2X Straight Bevel (Gear)** with the following properties.

Properties

Module:

2

Number of Teeth:

30

Pinion's No. of Teeth:

30

Pressure Angle:

14.5

Face Width:

12

Hub Diameter:

25

Mounting Distance:

150

Nominal Shaft Diameter:

10

Keyway:

None

Show Teeth:

30

2) Apply the following **Standard Mates** ![Mate] to one of the *Bevel gears*.
   a) Make the shaft of the *Bevel gear* **Concentric** with the shaft of the *Small gear* that is in contact with the *Rack*.
   b) Make the ends of the two shafts **Coincident** as shown in the figure.
   c) Make **Plane 1** of the *Bevel gear* **Coincident** with **Plane 1** of the *Small gear*. This will lock them together so that they will rotate as one.

3) **Sketch** on the **Right Plane** of the *Assembly* and draw and **Dimension** the **Line** shown.

4) **Exit** the **Sketch**.

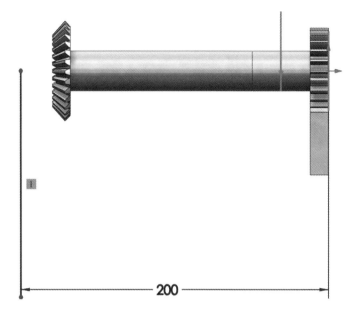

5) Apply the following **Standard Mates** to the other *Bevel gear*.

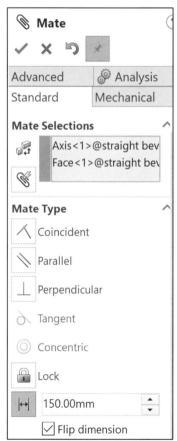

   a) Make the **axis** of the *Bevel gear* **Coincident** with the **line**. To get the axis to show up, highlight the inside of the hole and wait. The axis will show up after a while.

   b) Apply a **Distance mate** between the **Top Plane** and the **end of the** second *Bevel gear's* **shaft** of 150 mm.

6) Use a **Gear mate** with a **1:1** gear ratio to mate the *Bevel gears*.

   a) Align the gear teeth as shown in the figure.

   b) Select **Mate** .

   c) Select the *Mechanical* tab.

   d) Select the **Gear** mate.

   e) <u>Mate Selections:</u> Select the **inside** of the **holes** of both *Bevel gears*.

   f) Enter a gear ratio of **1:1**.

   g)

   h) Test the mate to see if it works. The mate may have to be *Reversed*.

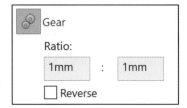

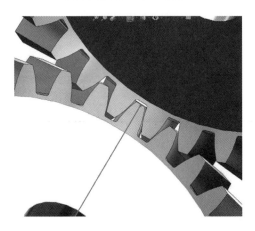

13.4.9) Inserting a New Part

1) In the *Assembly* tab, select **New Part** (under the *Insert Components* command).

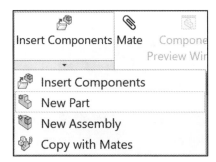

2) Select the end of the *Bevel gear's* shaft. This will be your sketch plane. Once done, the assembly will turn in to wire frame.

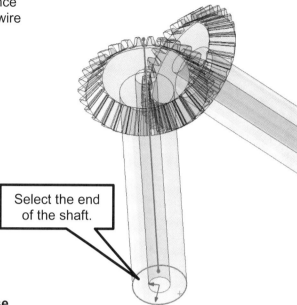

Select the end of the shaft.

3) **Sketch** and **Dimension** the **Ellipse** shown in the figure. Use **Ctrl + 8** to view the sketch plane straight on.

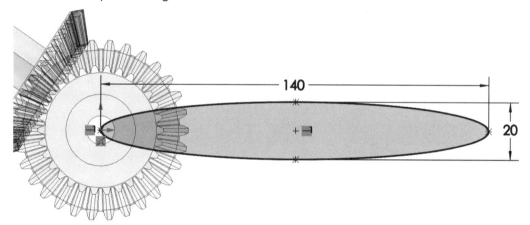

4) Use **Circular Sketch Pattern** to create a total of **4** ellipses using the center of the shaft as the rotation point.

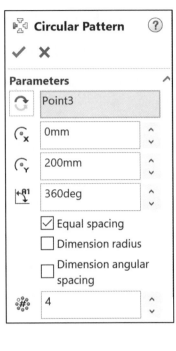

Circular Pattern

Parameters

Point3

0mm

200mm

360deg

☑ Equal spacing

☐ Dimension radius

☐ Dimension angular spacing

4

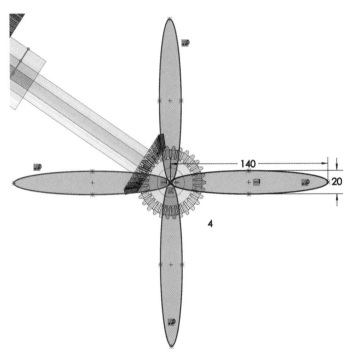

5) Zoom into the center of the ellipses. Note that they overlap. Use the Trim command to produce the sketch on the right.

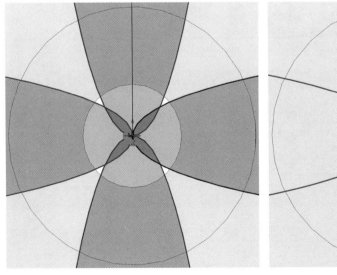

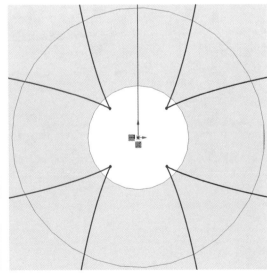

6) **Extrude** the sketch **10 mm**.

7) **Save As** your part as **FAN**.

8) **Right-Click** on the new part in the Feature Tree and select **Edit Assembly**.

9) Rotate the gear assembly. Note that the Fan rotates with the gears. An *InPlace* ![InPlace2] mate was automatically made.

10) **Save**

<u>NOTES:</u>

ADVANCED ASSEMBLIES IN SOLIDWORKS® PROBLEMS

P13-1) Use SOLIDWORKS® to create an assembly model of the *Milling Jack* shown. Use the appropriate tolerances and mates.

<u>Milling Jack</u>

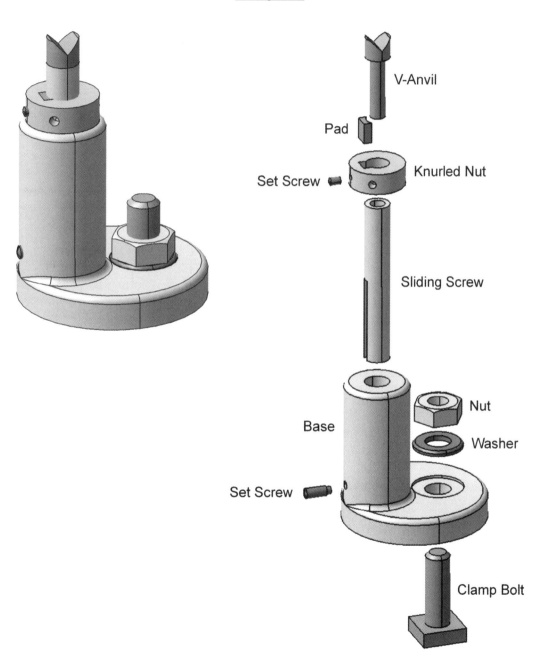

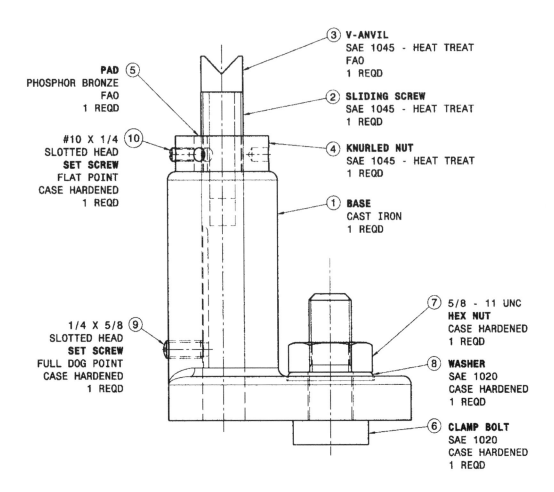

PAD ⑤
PHOSPHOR BRONZE
FAO
1 REQD

③ **V-ANVIL**
SAE 1045 - HEAT TREAT
FAO
1 REQD

② **SLIDING SCREW**
SAE 1045 - HEAT TREAT
1 REQD

#10 X 1/4 ⑩
SLOTTED HEAD
SET SCREW
FLAT POINT
CASE HARDENED
1 REQD

④ **KNURLED NUT**
SAE 1045 - HEAT TREAT
1 REQD

① **BASE**
CAST IRON
1 REQD

1/4 X 5/8 ⑨
SLOTTED HEAD
SET SCREW
FULL DOG POINT
CASE HARDENED
1 REQD

⑦ 5/8 - 11 UNC
HEX NUT
CASE HARDENED
1 REQD

⑧ **WASHER**
SAE 1020
CASE HARDENED
1 REQD

⑥ **CLAMP BOLT**
SAE 1020
CASE HARDENED
1 REQD

Base

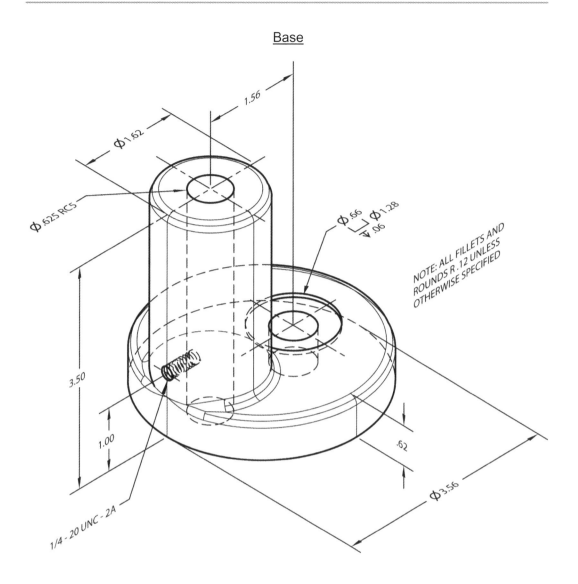

1.56

$\emptyset$1.62

$\emptyset$.625 RC5

$\emptyset$.56 $\emptyset$1.28
.06

NOTE: ALL FILLETS AND
ROUNDS R .12 UNLESS
OTHERWISE SPECIFIED

3.50

1.00

.62

$\emptyset$3.56

1/4 - 20 UNC - 2A

Sliding Screw

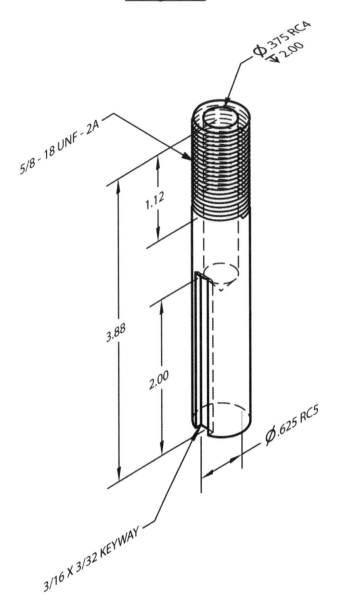

V-Anvil

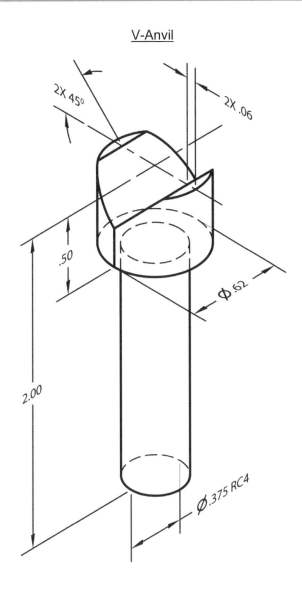

Knurled Nut

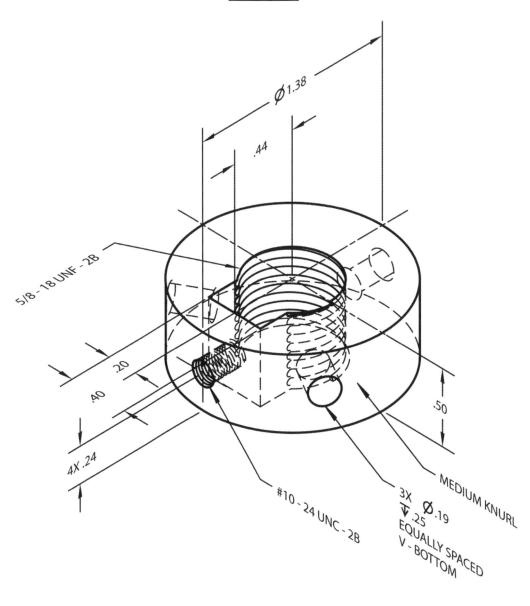

Pad

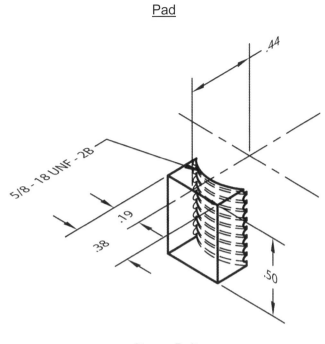

Clamp Bolt

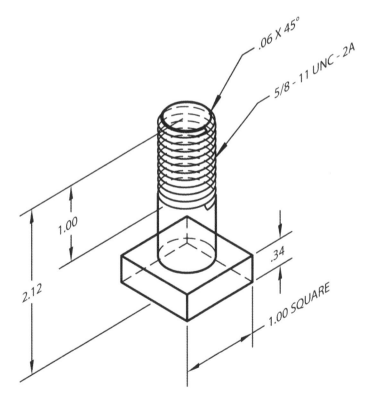

Washer

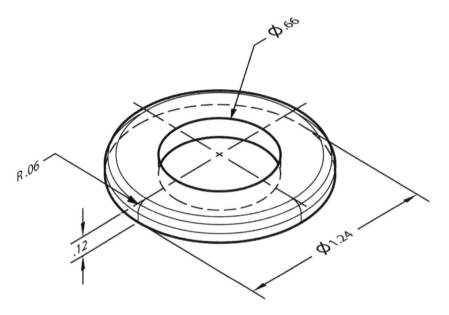

CHAPTER 14

3D SKETCHES IN SOLIDWORKS®

CHAPTER OUTLINE

14.1) COMMAND MANAGER 2

14.2) SKETCH 4

    14.2.1) Spline 4

    14.2.2) 3D Sketch 6

14.3) HANDLEBAR TUTORIAL 7

    14.3.1) Prerequisites 7

    14.3.2) What you will learn 7

    14.3.3) Top bar 9

    14.3.4) Spline 13

    14.3.5) Drop-down bar 17

3D SKETCHES IN SOLIDWORKS® PROBLEMS 19

CHAPTER SUMMARY

In this chapter, you will learn how to create realistic assemblies in SOLIDWORKS®. SOLIDWORKS® has many different types of mates that can be used to assemble parts. This chapter will focus on both standard and advanced mates. By the end of this chapter, you will be able to create assemblies with standard and advanced mates to produce assemblies that behave in a realistic manner.

14.1) COMMAND MANAGER

The *Command Manager* (shown in Figure 14.1-1) is the ribbon of commands at the top of the modeling area. Sometimes a command is stacked under another command or is not available within a tab and is inconvenient to access. SOLIDWORKS® gives you the ability to add commands to the *Command Manager* to increase your workflow efficiency.

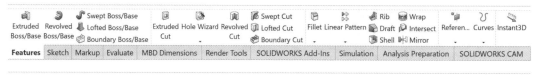

Figure 14.1-1: Command Manager

14.1.1) Adding commands to the *Command Manager*

The *Command Manager* may be customized to hold any command that you find useful. To add or remove commands, follow these steps.

Adding commands to the Command Manager

1) **Right click** on the *Command Manager* in an open area (an area with no commands) and select **Customize**.
2) In the *Customize* window, select the **Commands** tab. See Figure 14.1-2.
3) Select the *Category* in which the command resides. A set of *Buttons* will appear on the right side.
4) **Click and drag** the desired command to the desired location in the *Command Manager*.

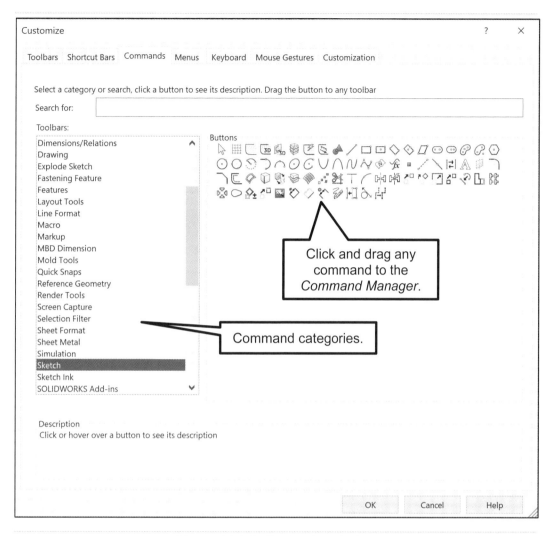

Figure 14.1-2: *Customize* window

14.2) SKETCH

In Chapter 1, we discussed that before a solid can be created, a two-dimensional sketch that will define the solid's shape must first be created. This chapter will expand that knowledge to using three-dimensional sketches. Many different sketch elements in the **Sketch tab** shown in Figure 14.2-1 can be used to create both 2D and 3D sketches. In this chapter, we will learn how to use **Splines** and other more familiar sketch elements to create three-dimensional sketches.

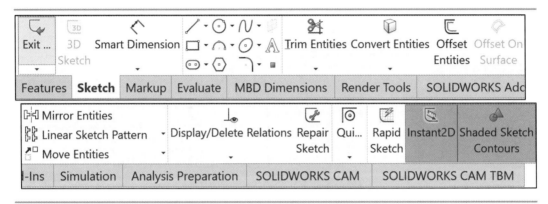

Figure 14.2-1: *Sketch* tab commands

14.2.1) Spline

A **Spline** is a fitted curve controlled by points. In general, a simple *Spline* is created by clicking at different locations until a fitted curve is created. Points are defined by the locations clicked. The location of these points can be edited after the *Spline* has been created. There are four different ways a spline can be created. Use the steps listed below to edit the point locations and tangencies associated with a *Spline*.

- **Spline:** $\boxed{\mathsf{N} \quad \text{Spline}}$: A fitted curve controlled by points.

- **Style Spline:** $\boxed{\mathsf{N} \quad \text{Style Spline}}$ Used as a bridge curve between two existing entities.

- **Spline on Surface:** $\boxed{\text{\textcircled{\tiny 0}} \quad \text{Spline on Surface}}$ Sketch a spline on an existing surface.

- **Equation Driven Curve:** $\boxed{\text{\tiny fx} \quad \text{Equation Driven Curve}}$ Allows the creation of a *Spline* using equations for the x, y, and z-coordinates.

Editing a *Spline*

1) Click on a *Point* and enter the (x,y,z) coordinate. Note: Sketch relations may also be added to points. See Figure 14.2-2.
2) Each point also has a *Spline handle*. These handles control tangency. The *Spline handles* may be rotated and lengthened to achieve the desired effect.

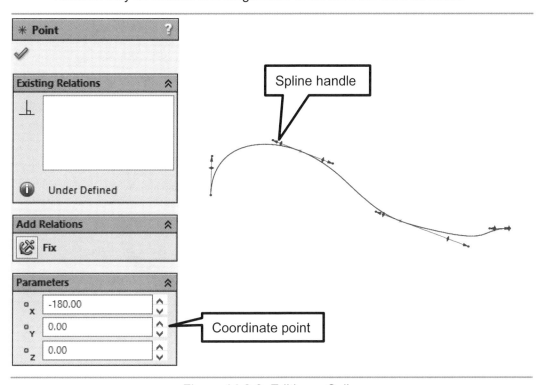

Figure 14.2-2: Editing a *Spline*

14.2.2) 3D Sketch

The **3D Sketch** command is located in the *Sketch* tab. A *3D Sketch* is a non-planar sketch that may be used as a guide curve in the *Sweep* or *Loft* commands. A *3D Sketch* may be drawn on the **XY, YZ** or **ZX** plane. You can even switch between the planes within a single sketch. For example, a drawing can be started on the **XY** plane and then can be switched to drawing on the **YZ** plane. To switch between planes, select the **TAB** key. Figure 14.2-3 shows an example of how a 3D sketch was used along with the Swept command to create a part.

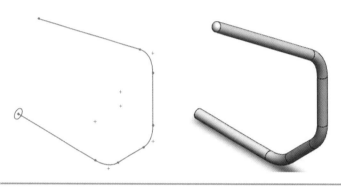

Figure 14.2-3: 3D Sketch and Swept command

14.3) HANDLEBAR TUTORIAL

14.3.1) Prerequisites

Before starting this tutorial, you should have completed the listed tutorials.

- Chapter 1 – Connecting Rod Tutorial
- Chapter 9 – Microphone Base Tutorial
- Chapter 9 – Microphone Arm Tutorial
- Chapter 9 – Boat Tutorial

14.3.2) What you will learn

The objective of this tutorial is to introduce you to the SOLIDWORKS'® 3D Sketching capabilities. You will be modeling the *Drop-Down Bicycle Handlebar* shown in Figure 14.3-1. Specifically, you will be learning the following commands and concepts.

View

- Adding commands to the Command Manager

Sketch

- 3D Sketch
- Spline

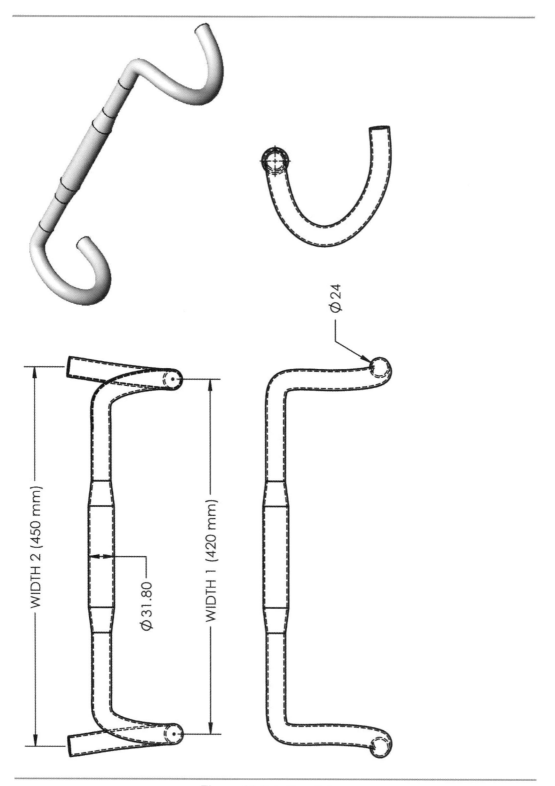

Figure 14.3-1: Handlebar

14.3.3) Top bar

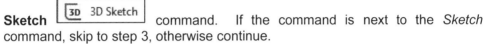

1) Open a **New part** 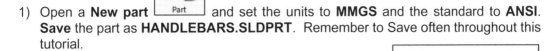 and set the units to **MMGS** and the standard to **ANSI**. **Save** the part as **HANDLEBARS.SLDPRT**. Remember to Save often throughout this tutorial.

2) **Add** the **3D Sketch** command to the **Command Manager** if it is not already there.
   a) Click on the **Sketch** tab.
   b) Underneath or next to the *Sketch* command is the **3D Sketch** | 3D 3D Sketch | command. If the command is next to the *Sketch* command, skip to step 3, otherwise continue.
   c) **Right click** on the *Command Manager* in an open area (an area with no commands) and select **Customize**.
   d) In the *Customize* window, select the **Commands** tab. See Figure 14.1-2.
   e) On the left, select **Sketch**.
   f) On the right, **Click and drag** next to the *Sketch* command in the *Command Manager*.

➤ To learn more about ***Adding Commands to the Command Manager*** see section 14.1.1.

---

3) Select the **3D Sketch** | 3D 3D Sketch | command. Take note of the coordinate axis and the directions that each axis points.

➤ To learn more about ***3D Sketches*** see section 14.2.2.

---

4) Select the **Line** | / Line | command. Notice that the cursor changes and shows either **XY, YZ** or **ZX**. This indicates on which plane the drawing will begin. Hit the **TAB** key and cycle through the planes. Then stop when the cursor reads **XY**. Start the *Line* at the **origin** and have it travel in the **X-direction** for approximately **60 mm**.

5) **Dimension** the *Line* and make it **60 mm** long.

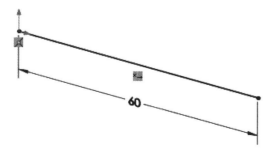

6) **Exit** the sketch by selecting the **3D Sketch** 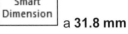 command.

7) **Sketch** on the **Right plane,** then draw and **Dimension** a **31.8 mm** diameter **Circle** that is **coincident** with the origin.

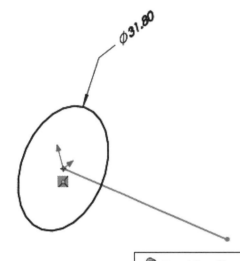

8) **Swept Boss/Base** the *Circle* along the *Line*. If the *Swept* command is grayed out, **Rebuild** .

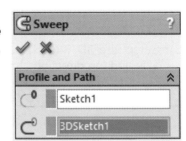

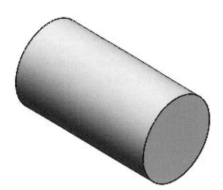

9) Start a new **3D Sketch** 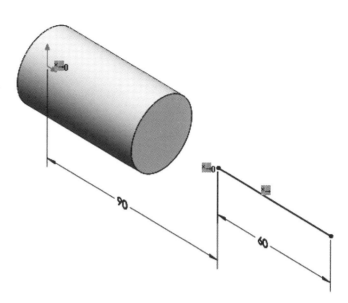 and draw a **Line** on the **XY** plane and in the **x-direction** with the following dimensions. Don't worry about lining it up exactly with the origin. We will take care of that in the next step.

10) **Exit** the *Smart Dimension* command and then select the origin, hold the **Ctrl** key and then select the end of the line that was just created. Constrain it to be **Along X**. Notice that sketch relation symbols. If the sketch relations cannot be seen, select **View – Hide/Show – Sketch Relations**.

11) **Exit** the **3D Sketch**.

12) Create a reference **Plane** 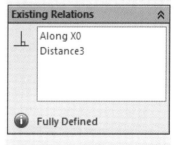 that is **90 mm** to the right of the **Right plane**.

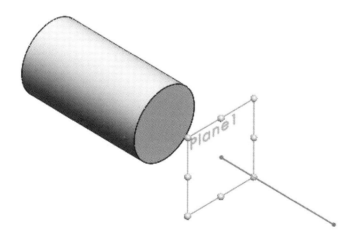

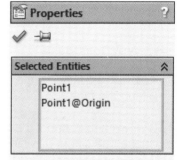

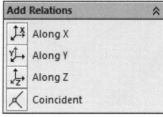

13) On the new plane, draw a **24 mm** diameter **Circle** ⊘ | Circle that is **coincident** with the origin.

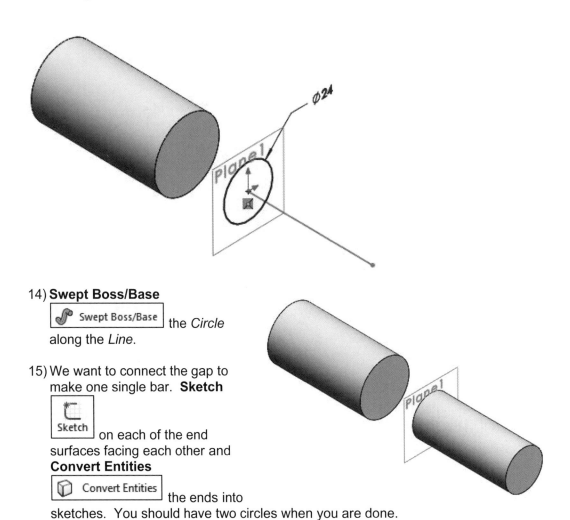

14) **Swept Boss/Base**

🐛 Swept Boss/Base the *Circle* along the *Line*.

15) We want to connect the gap to make one single bar. **Sketch**

⌐ Sketch on each of the end surfaces facing each other and **Convert Entities**

⬚ Convert Entities the ends into sketches. You should have two circles when you are done.

16) **Loft Boss/Base** | ⬇ Lofted Boss/Base | between each pair of circles. If the *Loft* is twisted, move the green dots so that they line up. (A description of the *Loft* command is given in the *Microphone Base and Arm Tutorials.*)

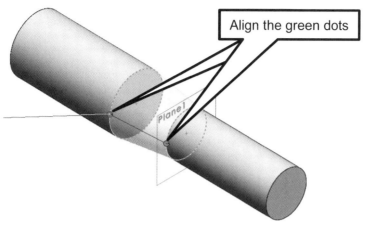

Align the green dots

14.3.4) Spline

1) View the part from the **Top**. The **Shift + arrow** key will rotate the part so that it can be moved into the position shown. The axis should show the **ZX** plane.

2) Start a new **3D Sketch** | 3D Sketch | and draw a 6-point **Spline** | ∿ Spline | on the **ZX** plane that looks similar to what is shown. **Double click** on the last point to end the *Spline*.

> ➢ To learn more about *Splines* see section 14.2.1.

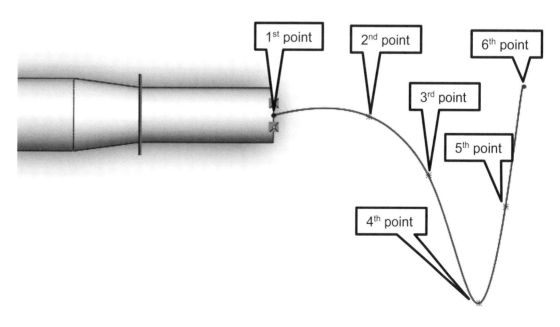

1st point   2nd point   6th point   3rd point   5th point   4th point

3) Get out of the *Spline* command and then **click on *Point 1***
of the *Spline*. In the *Point* window, enter the following
coordinates.
- First point = **150,0,0**

4) Adjust the coordinates of the other points; the sketch will
look like the figure.
   a. Second point = **180,0,0**
   b. Third point = **210,-19,65**
   c. Fourth point = **210,-60,89**
   d. Fifth point = **217,-127,15**
   e. Sixth point = **225,-127,-40**

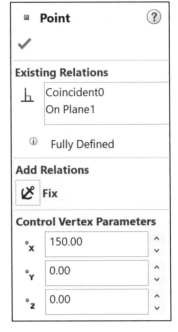

	Point	?
✓		

Existing Relations

└	Coincident0
	On Plane1

ⓘ   Fully Defined

Add Relations

⚓ **Fix**

Control Vertex Parameters

•X	150.00	⌄
•Y	0.00	⌄
•Z	0.00	⌄

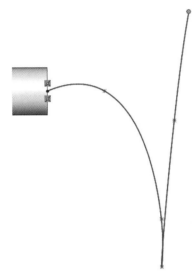

5) Show the isometric view (**Ctrl + 7**). This is
now the general shape of the drop-down part of the
handlebar, but we need to tweak the tangencies.

6) **View** the handlebar from the **top** and click on the **first point**. Notice that a spline handle appears. Drag the end of the arrow so that it is horizontal. Then, adjust the tangency of the **second point** to create an approximate straight line between the first and second points.

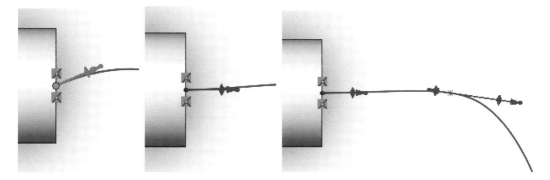

7) In the same view, adjust the tangency of the **sixth point** to create an approximately straight line between the fifth and sixth point.

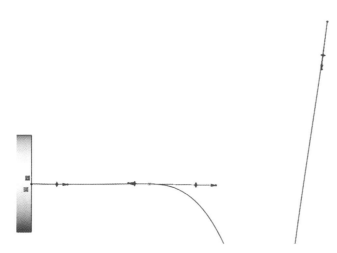

8) **View** the handlebar from the **right side**. Adjust the tangency of the **sixth and fifth points** to create an approximately straight line between them.

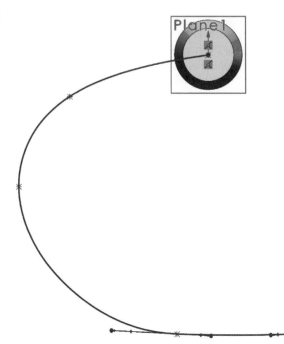

9) When done, the shape should be similar to what is shown. **Exit** the **3D Sketch**.

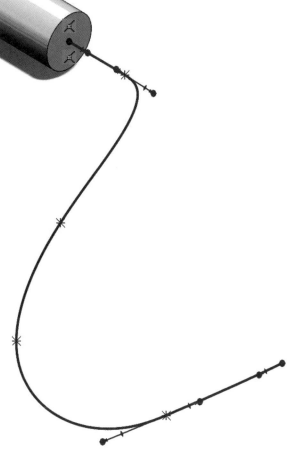

14.3.5) Drop-down bar

1) **Sketch** on the right end of the top bar, **Convert Entities** Convert Entities , and **Sweep** Swept Boss/Base this sketch along the *Spline*.

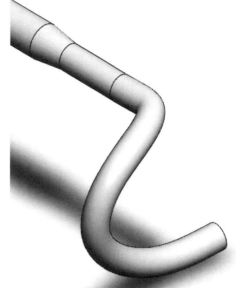

2) **Mirror** Mirror the <u>handlebar</u> about the **Right plane** to create the other half. Notice that you get an error. This is because you will need to mirror each element separately.

3) **Mirror** Mirror each component of the handlebar separately about the **Right plane** to create the other half.

4) **Shell** 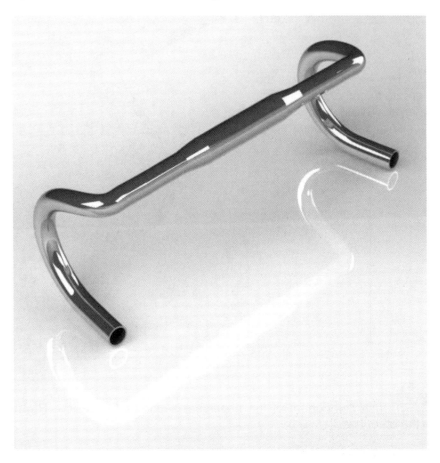 the part to **2 mm** eliminating the two end faces. (The *Shell* command is described in *Microphone Tutorial.*)

5) Make the handlebar **1060 Alloy** Aluminum.

> ➤ **Error?** If you get a *Rebuild Error* do the following.
> - Select **OK** in the error window.
> - At the bottom of the *Shell* option window there will appear an *Error Diagnostics* area. Click the ***Entire body*** radio button and then select **Check body/faces**.
> - It will show the problem area on the model.
> - Go in and fix that area and try the Shell again.

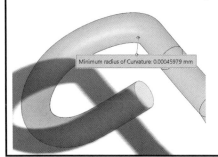

Minimum radius of Curvature: 0.00045979 mm

3D SKETCHES IN SOLIDWORKS® PROBLEMS

P14-1) Model one of the following objects.

a) A whisk

b) Spring – Use at least one SWEEP command.

c) An exhaust pipe

d) A fluorescent light bulb

e) An oven rack

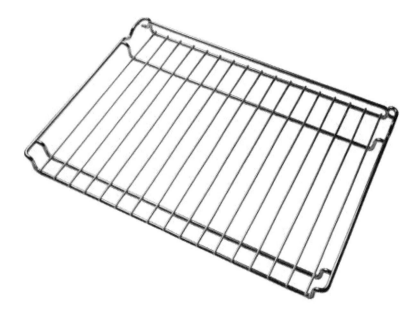

CHAPTER 15

RENDERING IN SOLIDWORKS®

CHAPTER OUTLINE

15.1) WHAT IS A RENDERING? 2

15.2) APPEARANCE, SCENES, AND DECALS 3

15.2.1) Appearance 3

15.2.2) Scene 4

15.2.3) Lighting 5

15.2.4) Decals 5

15.2.5) Camera 5

15.2.6) Display Manager 5

15.3) RENDER TOOLS 5

15.3.1) PhotoView 360 5

15.3.2) Render Tools 7

15.3.3) Options 7

15.3.4) Preview Window 9

15.3.5) Render Region 10

15.3.6) Final Render 10

15.4) GLASS VASE TUTORIAL 11

15.4.1) Prerequisites 11

15.4.2) What you will learn? 11

15.4.3) The Vase 12

15.4.4) Appearance 15

15.4.5) PhotoView 360 17

15.5) LIGHT BULB TUTORIAL 22

15.5.1) Prerequisites 22

15.5.2) What you will learn? 22

15.5.3) Appearance 23

15.5.4) PhotoView 360 25

RENDERING IN SOLIDWORKS® PROBLEMS 33

CHAPTER SUMMARY

In this chapter, you will learn how to create renderings in SOLIDWORKS®. A rendering is a photorealistic image of a 3D model. By the end of this chapter, you will be able to create simple renderings of a part or assembly.

15.1) WHAT IS A RENDERING?

A rendering is a photorealistic image of a part or assembly. In SOLIDWORKS® you can apply predefined appearances to your model and render it so that it looks more realistic. Figure 15.1-1 shows a glass vase. The picture on the left is a model of the glass vase with glass applied as the material. The picture on the right is the same glass vase with glass applied as the appearance and a reflective floor used as the scene. There are many predefined appearances and scenes from which to choose.

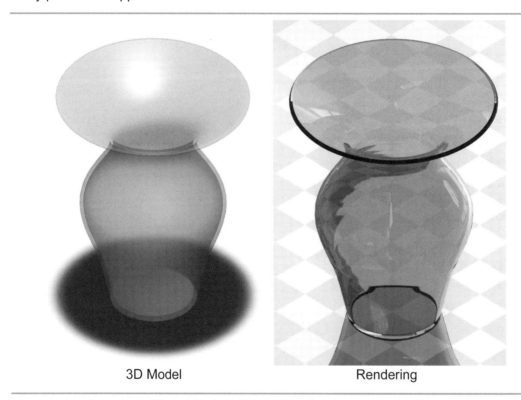

3D Model   Rendering

Figure 15.1-1: Rendering of a glass vase

15.2) APPEARANCE, SCENES, AND DECALS

15.2.1) Appearance

An appearance defines the visual properties of a model, including color and texture. Appearances do not affect the physical properties of the model. The physical properties are defined by the assigned materials.

Appearances can be applied to an object through the *Task Pane*. SOLIDWORKS® has many predefined appearances. To apply an appearance, use the following steps.

1) Click on **Appearances, Scenes, and Decals** in the *Task Pane*.
2) Expand Appearances (see Figure 15.2-1).
3) Expand the material or characteristic.
4) Click on the category.
5) Click on the appearance you wish to apply and drag it to the part. Once you drag the appearance to the part you will get a choice to apply it to *Face*, *Feature*, *Body*, or *Part*. 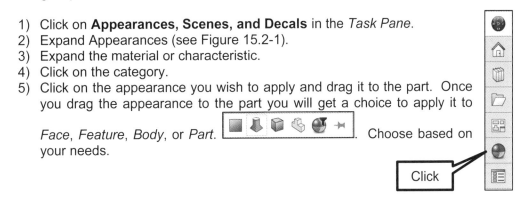. Choose based on your needs.

Click

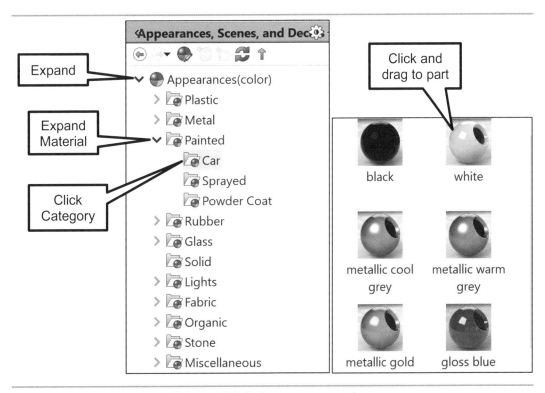

Figure 15.2-1: Appearance options

15.2.2) Scene

Scenes provide a visual backdrop behind a model. With PhotoView 360, scenes provide a realistic light source, including illumination and reflections, requiring less manipulation of lighting. The objects and lights in a scene can form reflections on the model and can cast shadows on the floor.

Scenes can be applied through the *Task Pane*. SOLIDWORKS® has many predefined scenes. To apply a scene, use the following steps.

1) Click on **Appearances, Scenes, and Decals** in the *Task Pane*.
2) Expand scenes (see Figure 15.2-2).
3) Expand the scene library.
4) Click on the category.
5) Double-click on the scene you wish to apply.

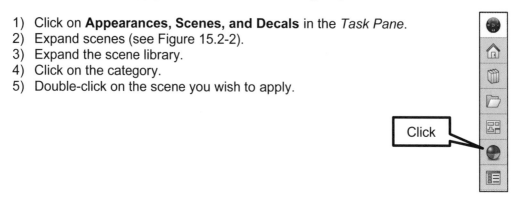

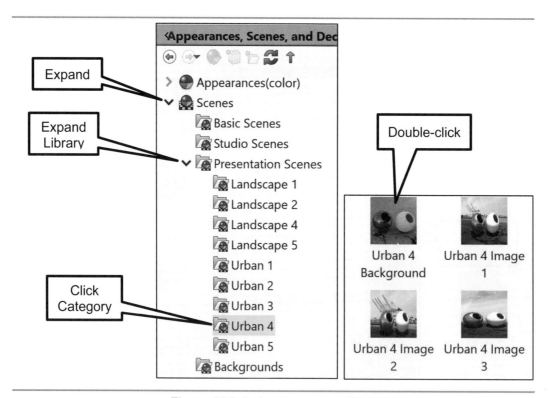

Figure 15.2-2: Appearance options

15.2.3) Lighting

The direction, intensity, and color of light that is illuminating the model can be adjusted. Light sources of various types can be added, and their characteristics can be modified to illuminate the model as needed. The predefined lighting of a scene can also be edited.

15.2.4) Decals

A decal is a 2D image applied to a model. Decals can be used to apply company names or instructions to models. They also can be used to apply images that substitute for details in your model. For example, keys on a computer keyboard.

15.2.5) Camera

Cameras can be added to a model document and viewed from the camera's perspective.

15.2.6) Display Manager

The *Display Manager* is akin to the *Feature Manager Design Tree*. The *Display Manager* shows all the appearances that have been applied, the scene, decals, light settings, and cameras added. Figure 15.2-3 shows the *Display Manager*.

15.3) RENDER TOOLS

15.3.1) PhotoView 360

PhotoView 360 is a SOLIDWORKS® add-in that produces photo-realistic renderings of SOLIDWORKS® models. The rendered image incorporates the appearance, lighting, scene, and decals included with the model. By default, PhotoView 360 is not active when you start SOLIDWORKS. To activate it and thus activate the *Render Tools* tab, you need to add it in. To add-in PhotoView 360 use the following steps.

1) Select **Tool – Add-Ins…** from the pull-down menu.
2) Check the checkbox next to PhotoView 360 in the *Add-Ins* window. The left check box will activate it for the current session, the right checkbox will activate it every time you start SOLIDWORKS.

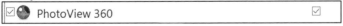

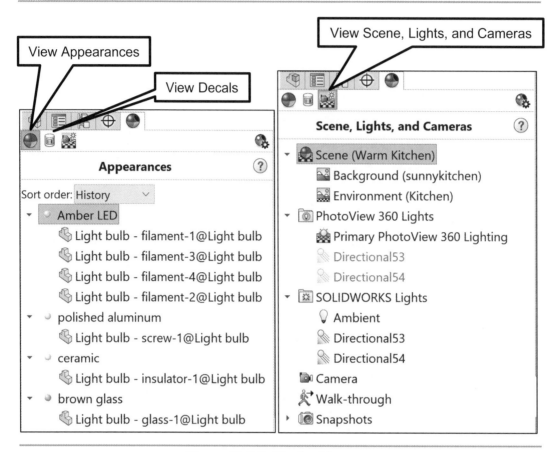

Figure 15.2-3: Display Manager

15.3.2) Render Tools

Once appearances and a scene have been applied, it is time to render the object. The *Render Tools* commands are located in the **Render Tools tab** shown in Figure 15.3-1.

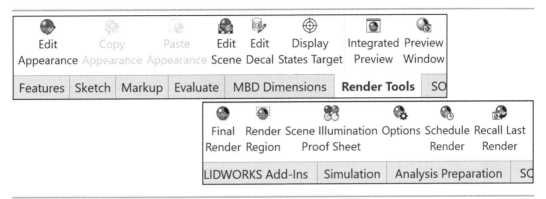

Figure 15.3-1: Render Tools tab commands

15.3.3) Options

Before the object is rendered, the **Options** [Options] should be set. The options window is shown in Figure 15.3-2. A few important settings to consider are listed here.

- **Output Image Settings:** This is the resulting rendered image size. The larger the setting, the longer it will take to render the object.
- **Image Format:** Different image formats such as JPEG and TIF can be chosen.
- **Preview render quality:** Set the preview render to **Good** so that it doesn't take a long time to refresh when changes are made.
- **Final render quality:** This quality together with the image size will depend on the final use of the image.
- **Bloom:** Bloom is used if there is a light source within the object.

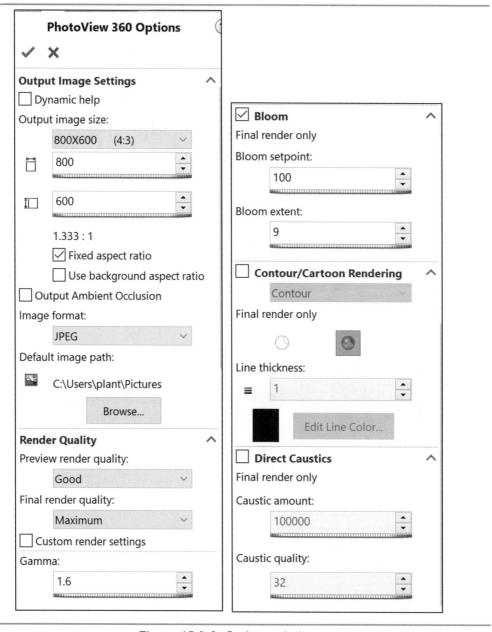

Figure 15.3-2: Options window

15.3.4) Preview Window

The **Preview Window** 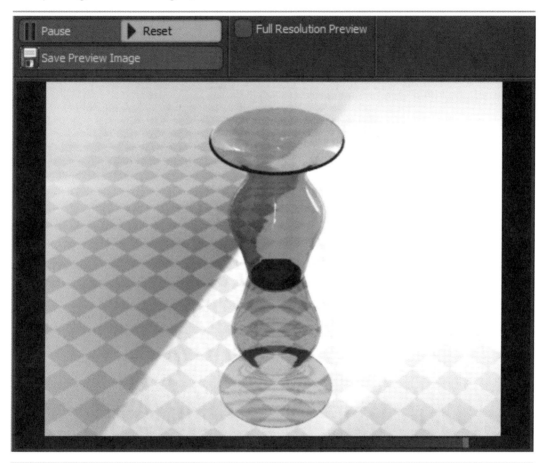 allows for a preview of the render to be seen so that adjustments can be made, if necessary, before performing the final render. Figure 15.3-3 shows an example of the *Preview Window*. The *Preview Window* can be **Paused** from updating if it is slowing down the computer's performance. If the *Preview Window* is not updating with the changes, the **Reset** button can be selected.

Figure 15.3-3: Preview Window

15.3.5) Render Region

A **Render Region** can be selected if the entire visible area is not wanted to be rendered.

15.3.6) Final Render

Once satisfied with what is shown in the *Preview Window*, the **Final Render** can be created. The final render may take a long time to complete. It depends on the size of the image and the speed of the computer. When the final render is completed, the image can be saved.

15.4) GLASS VASE TUTORIAL

15.4.1) Prerequisites

Before starting this tutorial, you should have completed the following tutorial.

Prerequisite Tutorials

- Chapter 9 – Microphone Tutorial

15.4.2) What you will learn?

The objective of this tutorial is to introduce you to the SOLIDWORKS'® ability to create a realistic looking part. You will be modeling the *Glass Vase* shown in Figure 15.4-1 and then creating a rendering of the vase that looks very realistic. Specifically, you will be learning the following commands and concepts.

Render Tools

- PhotoView 360
- Appearance
- Scene
- Add Camera
- Preview Window
- Final Render

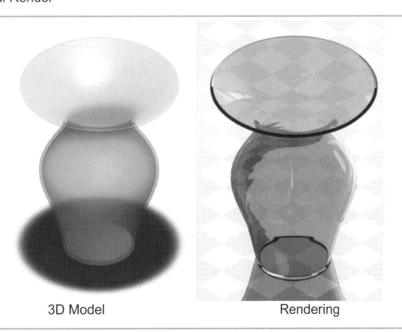

| 3D Model | Rendering |

Figure 15.4-1: Rendering of a glass vase

15.4.3) The Vase

1) **Start SOLIDWORKS** and start a **new part** .

2) Set the drafting standard to **ANSI** and set the text to **upper case** for notes, tables, and dimensions. (*Options* ⚙ *– Document Properties – Drafting Standard*)

3) Set the units to **IPS** (millimeters, grams, second) and the **decimal = 0.12**. (*Options* ⚙ *– Document Properties – Units*)

4) Save the part as **GLASS VASE.SLDPRT** (**File – Save**). Remember to save often throughout this project.

5) Create 4 Reference **Planes** [📄 Plane] that are parallel to the **Top Plane** and offset by **2.00 inches**.

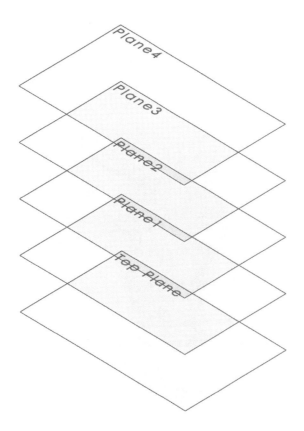

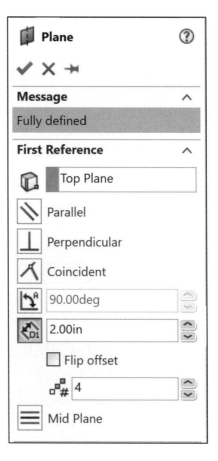

6) Draw a **Circle** on each of the planes, starting with the *Top Plane*, centered at the origin.

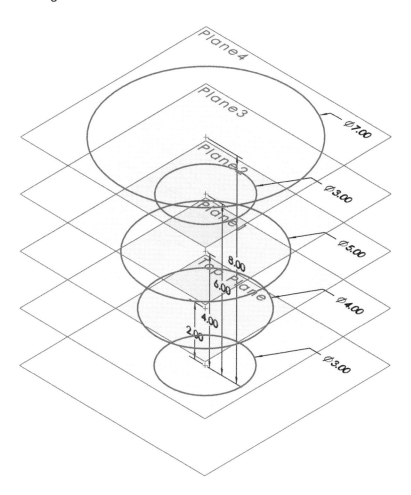

7) **Loft** Lofted Boss/Base the circles. Make sure that the green dots are all aligned.

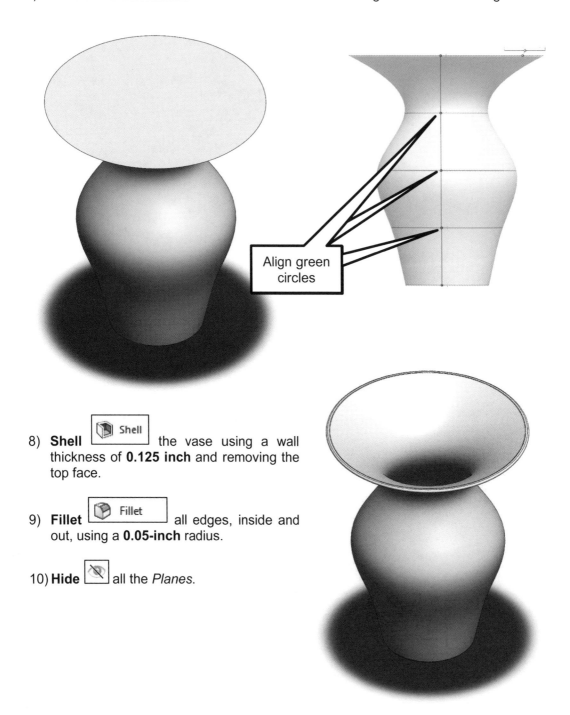

Align green circles

8) **Shell** Shell the vase using a wall thickness of **0.125 inch** and removing the top face.

9) **Fillet** Fillet all edges, inside and out, using a **0.05-inch** radius.

10) **Hide** all the *Planes*.

15.4.4) Appearance

1) Make the *Vase's* **Appearance** be **Green Glossy Glass**.
    a) Click on **Appearances, Scenes, and Decals** in the *Task Pane*.
    b) Expand *Appearances*.
    c) Expand *Glass*.
    d) Click on **Gloss**.
    e) Click on **green glass** and **drag it to the part**. A pop-up window will appear that allows you to choose whether to apply it to the *Face*, *Loft*, *Body*, or *Part (Glass Vase)*. Choose **Glass Vase**.

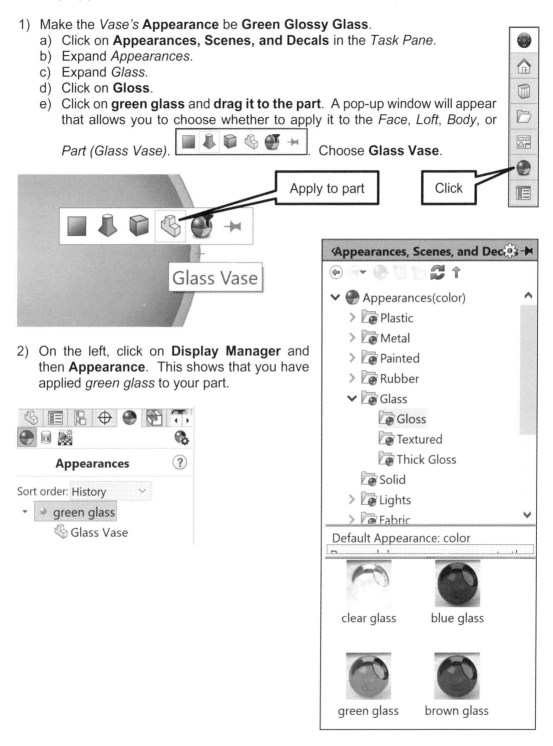

2) On the left, click on **Display Manager** and then **Appearance**. This shows that you have applied *green glass* to your part.

3) Choose a **Reflective Checkered Floor** for the *Scene*.
   a) Click on **Appearances, Scenes, and Decals** in the *Task Pane*.
   b) Expand *Scenes*.
   c) Click on **Studio Scenes**.
   d) Double-click on **Reflective Floor Checkered**.

4) On the left, click on **Display Manager** and then **Scene, Lights, and Cameras**. This shows that you have applied *Reflective Floor Checkered* to your model.

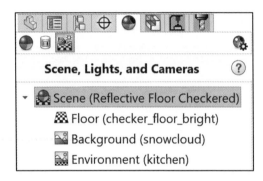

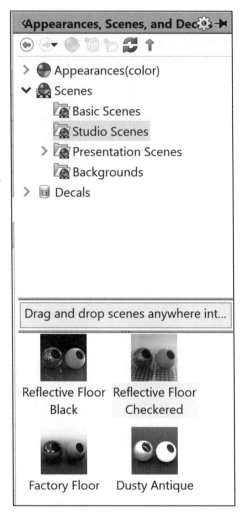

15.4.5) PhotoView 360

1) Add-In *PhotoView 360* if it is not already active. In the pull-down menu, click on **Tools – Add-Ins**. Check the boxes to activate the *PhotoView 360* add-in. Or click on **PhotoView 360**  in the *SOLIDWORKS Add-Ins* tab.

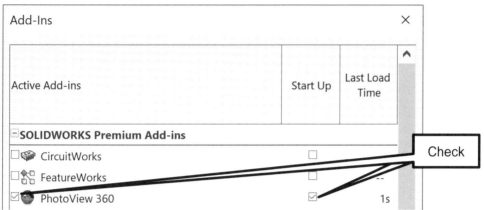

2) A tab should appear called **Render Tools**. Click on it and then click on **Options** [Options]. Use the following settings.
    a) Output image size: **800x600 (4:3)** (Note: the larger this is the slower the final render will take.)
    b) Image format: **JPEG**
    c) Render Quality:
        i.   Preview render quality: **Good** (This allows the preview window to update more quickly.)
        ii. Final render quality: **Maximum**
    d) Gamma: **1.6**
    e) Bloom: **Deselect**
    f) ✅

3) Click on it and then click on **Preview Window** 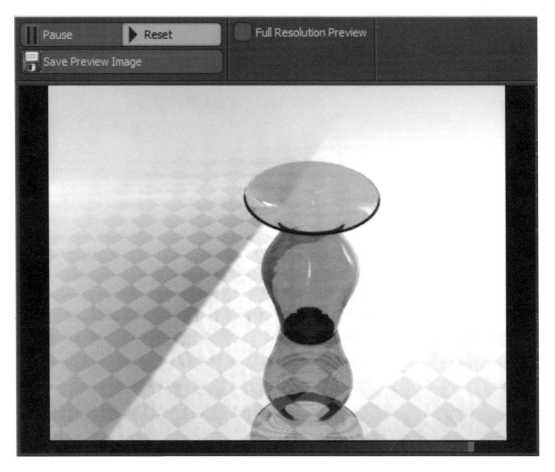. A window should appear saying that you don't have a camera. We don't want to add a camera yet, so select **Turn on Perspective View**. The time it takes the preview window to show an image depends on the speed of your computer.

4) If you don't like the way the vase is positioned, it can be moved around and rotated. The preview window will show what is being done. If the preview window is not tracking the changes, select the *Reset* button.

Final Render

5) Once satisfied with the image in the preview window, select **Final Render** . The final render may take a bit depending on the speed of the computer.

6) Once the final render is completed, select **Save Image**.

7) Access the *Preview Window* and close it.

8) Remove the perspective view by clicking from the pull-down menu, **View-Display-Perspective**. If the program did not ask you to turn on the perspective view, you will not have to turn it off.

9) On the left, click on **Display Manager** and then **Scene, Lights, and Cameras**. **Right click** on *Camera* and select **Add Camera**.

10) The screen will split. On the left screen, zoom out until you can see the camera. The right screen will show you the camera's view.

Scene, Lights, and Cameras

- Scene (Reflective Floor Ch
  - Floor (checker_floor_bri
  - Background (snowclou
  - Environment (kitchen)
- PhotoView 360 Lights
  - Primary PhotoView 360
  - Directional3
  - Directional4
- SOLIDWORKS Lights
  - Ca
  - W

  Add Camera
  Show Cameras
  Expand Item
  Collapse All
  Expand All
  Customize Menu

11) Click and drag the camera and watch the view on the right change. When you find an interesting view, bring up the **Preview Window**. You can always edit the camera if this is not what you like. When you like what you see, do a **Final render**.

15.5) LIGHT BULB TUTORIAL

15.5.1) Prerequisites

Before starting this tutorial, you should have completed the following tutorial.

Prerequisite Tutorials

- Glass Vase Tutorial

15.5.2) What you will learn?

The objective of this tutorial is to introduce you to the SOLIDWORKS'® ability to create realistic looking parts. You will be rendering the *Light Bulb* shown in Figure 15.5-1. Specifically, you will be learning the following commands and concepts.

Render Tools

- Appearance
- Scene
- Add Camera
- Preview Window
- Final Render
- Render Region

Figure 15.5-1: Light Bulb

15.5.3) Appearance

1) **Download** the following files and place them in a common folder.
   - **Light bulb – Student.SLDASM**
   - **Light bulb – filament.SLDPRT**
   - **Light bulb – glass.SLDPRT**
   - **Light bulb – insulator.SLDPRT**
   - **Light bulb – screw.SLDPRT**

2) **Open Light bulb – Student.SLDASM.** What you should see is a light bulb where everything is grey, but the glass on the light bulb is transparent.

3) Make the *Light bulb - glass* **Appearance** be **Brown Glossy Glass**.
   a) Click on **Appearances, Scenes, and Decals** in the *Task Pane* (on the right side).
   b) Expand *Appearances*.
   c) Expand *Glass*.
   d) Click on **Gloss**.
   e) Click on **brown glass** and drag it to the **Light bulb glass**. A pop-up window will appear that allows you to choose whether to apply it to the *Face, Feature, Body, Part, or Part@Assembly*. Choose **Light bulb – glass-1@Light bulb – Student.**

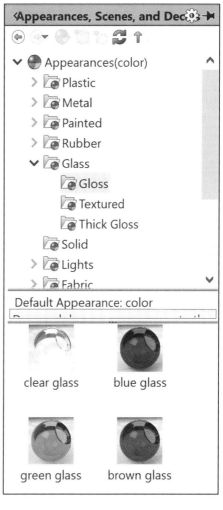

Apply to part within the assembly

Light bulb - glass-1@Light bulb - Student

4) On the left, click on **Display Manager** and then **Appearance**. This shows that you have applied *brown glass* to the *Light bulb - glass*.

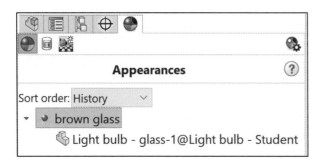

5) Click on **Appearance** 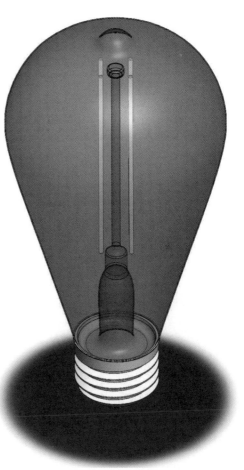 in the *Task Pane* (on the right side). Apply the following appearances to the rest of the parts.
   - Light bulb – Filament: **Lights – LED – Amber LED** (all four)
   - Light bulb – screw: **Metal – Aluminum – polished aluminum**
   - Light bulb – insulator: **Stone – Stoneware - ceramic**

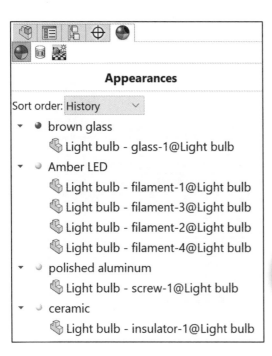

6) Choose **3 Point Beige** for the *Scene*.
   a) Click on **Appearances, Scenes, and Decals**  in the *Task Pane*.
   b) Expand *Scenes*.
   c) Click on **Basic Scenes**.
   d) Double-click on **3 Point Beige**.

7) On the left, click on **Display Manager** and then **Scene, Lights, and Cameras**. This shows that *Reflective Floor Checkered* has been applied to the model.

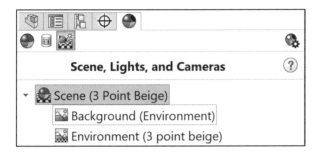

15.5.4) PhotoView 360

1) Add-In PhotoView 360 if it is not already active. In the *SOLIDWORKS Add-Ins* tab, click on **PhotoView 360** 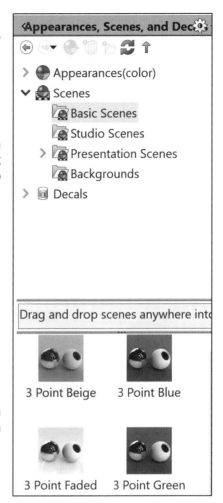.

2) A tab should appear called **Render Tools**. Click on it and then click on **Options**
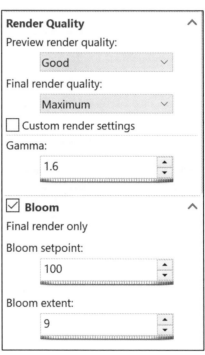
. Use the following settings.

   a) Output image size: **800x600 (4:3)** (Note: the larger this is, the slower the final render will take.)

   b) Image format: **JPEG**

   c) Render Quality:

      i. Preview render quality: **Good** (This allows the preview window to update more quickly.)

      ii. Final render quality: **Maximum**

   d) Gamma: **1.6**

   e) Bloom: **Activate** (This should be activated if any elements of light are being used.)

      i. Bloom setpoint: **100**

      ii. Bloom extent: **9**

   f)

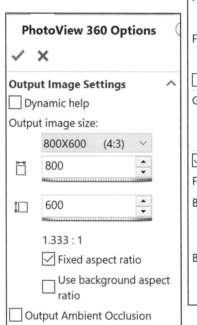

3) Click on it and then click on **Preview Window** 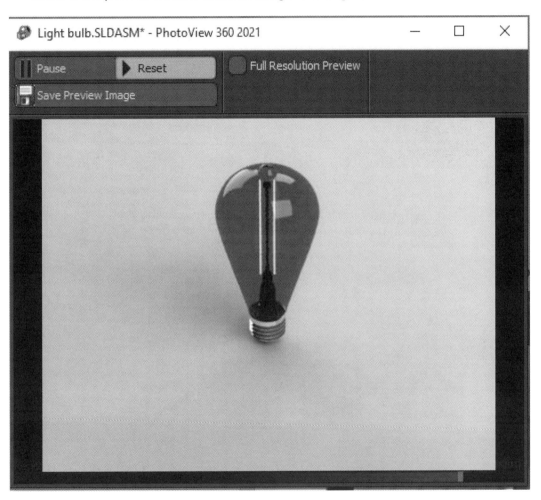 A window should appear saying that there is not a camera. We don't want to add a camera yet, so select **Turn on Perspective View**. The time it takes the preview window to show an image depends on the speed of the computer. If the positioning of the *Light bulb* is not ideal, it can be moved around and rotated. The preview window will show what is being done. If the preview window is not tracking the changes, select the Reset button.

4) Notice that the *brown glass* is too dark, so it needs to be lightened. On the left, click on **Display Manager** and then **Appearance**. Right-click on *brown glass* and select **Edit Appearance**.

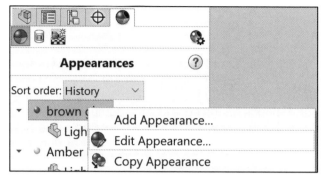

5) In the *brown glass* options window, click on the brown color box next to the eye dropper.

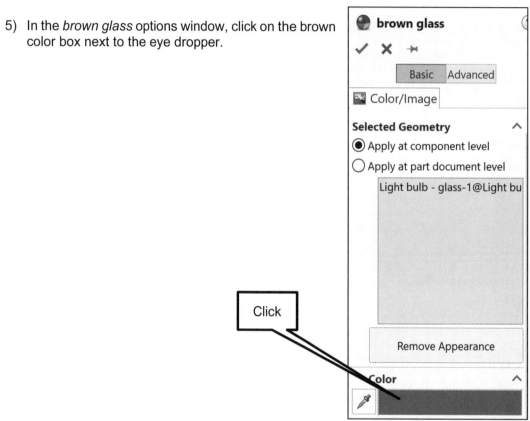

6) In the *Color* window, click on **Define Custom Colors>>**. Use the slider to lighten the brown color. When satisfied with the color choice, select **OK**.

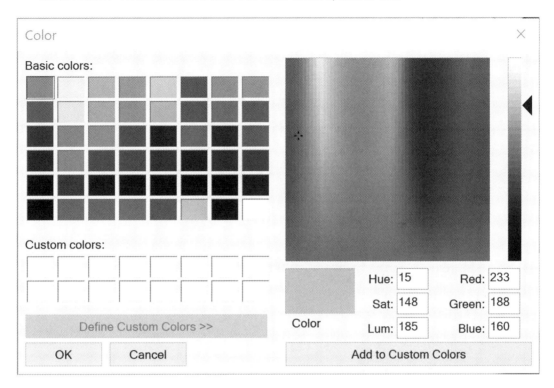

7) Click ☑ on the *brown glass* options window. Notice the change in the *Preview* window.

8) Once satisfied with the image in the preview window, select **Final Render** [Final Render]. The final render may take a bit of time depending on the speed of the computer. Once the final render is completed, select **Save Image**.

9) Access the *Preview Window* and close out of it.

10) View the part from the bottom. Hit the **space bar** to access the view cube and then select the panel indicated in the figure.

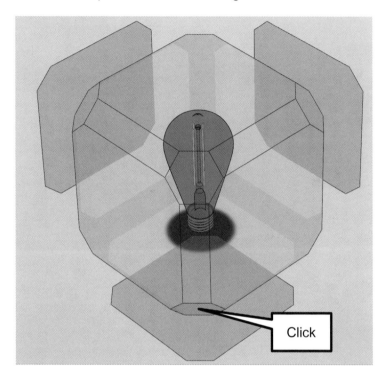

Click

11) Notice that the lights of the scene are not appealing and interfere with the view of the light bulb. Let's change the scene to something more pleasant. Click on **Appearance** in the *Task Pane* (on the right side). Expand **Scenes**, click on **Basic Scenes**, and then **double-click** on **Warm Kitchen**.

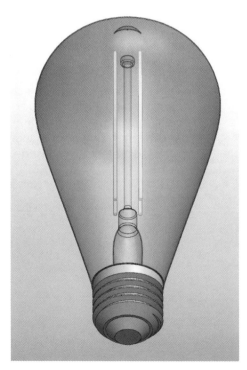

12) Click on **Preview Window** .

13) Click on **Render Region** . A pink box should appear. Adjust it to fit around the *Light bulb*. Look at the *Preview* window. **Close** the *Preview* Window and re-open it. Notice that what is shown on the screen and in the Preview window may be different. Adjust the **Render Region** so that all of the *Light bulb* can be seen on the screen and do a **Final Render** and **save** the image.

NOTES:

RENDERING IN SOLIDWORKS® PROBLEMS

P15-1) Model and render a pop bottle or can.

P15-2) Model and render a wine glass.

P15-3) Model and render a table.

P15-4) Model and render a flashlight.

P15-5) Model and render a hand tool.

P15-6) Model and render a re-usable water bottle.

NOTES:

APPENDIX A

USEFUL COMMANDS AND SHORTCUTS

A.1) VIEW COMMANDS

Command	Short-Cut Key, action, or status
Fit All	**F**
Pan	**Ctrl** + Press mouse wheel and move
Rotate	• **Ctrl + Arrow key** • Click mouse wheel and move
Rotate, 90°	**Shift + Arrow key**
Switch windows	**Ctrl + Tab**
View Cube	**Space Bar**
View, isometric	**Ctrl + 7**
View, normal	**Ctrl + 8**
Zoom	Scroll mouse wheel
Zoom in	**Z**
Zoom out	**z**

A.2) SKETCHING COMMANDS

Command	Short-Cut Key, action, or status
Copy	• **Ctrl** + click and drag • **Ctrl + c**
Editing sketch plane	Click on the sketch you wish to transfer to another plane, select the **Edit Sketch Plane** command, and then select the new plane or face.
Line, multiple	**click + release + click**
Line, single	**click + hold + drag**
Paste	**Ctrl + v**
Previous command	**s**
Reference planes, adding	Hold the **Ctrl** key down, click, and drag to create a reference plane.
Sketch relations, adding	Hold the **Ctrl** key down and select the entities that you wish to constrain. In the *Properties* window, select the relation you wish to apply.
Sketch relations, suppress	Press **Ctrl** while sketching
Sketch status	• Fully defined = **black** • Under defined = **blue** • Over defined = **red**

A.3) FEATURE COMMANDS

Command	Short-Cut Key, action, or status
Editing a feature	To edit the properties of a feature, click on the feature in the *Feature Manager Design Tree* and select **Edit Feature**
Rebuild	**Ctrl + b**

A.4) SELECTION COMMANDS

Command	Short-Cut Key, action, or status
Filters	o Edges = **e** o Faces = **x** o Vertices = **v** o Hide/Show all filters = **F5** o Off/On all selected filters = **F6**
Select, multiple	Hold **Ctrl** key and select

A.5) DRAWING COMMANDS

Command	Short-Cut Key, action, or status
Dimensions, copying to another view	**Ctrl** key + click and drag
Dimensions, moving to another view	**Shift** key + click and drag
Dimensions, moving within view	Hold the **Alt** key down when dragging dimensions to make them move smoothly.

A.6) ASSEMBLY COMMANDS

Command	Short-Cut Key, action, or status
Copy component	**Ctrl** + click and drag
Mate, smart	**Alt** + drag a component

APPENDIX B

INDEX

B.1) USER INTERFACE

Command or concept	Chapter
Adding commands to the command manager	14
Command manager	1
Feature manager design tree	1
Pull-down menu	1
Quick access toolbar	1
Quick unit switching	1
Status bar	1
Task pane	1
View (heads-up) toolbar	1

B.2) SETUP, FILE MANAGEMENT, AND HELP

Command or concept	Chapter
File types	1
Help	1
New drawing	2
New part	1
Options	1
Resources	1
Save	1
Units	1

B.3) SKETCH

Command or concept	Chapter
3D Sketch	14
Arcs	1, 9
Centerline	1
Centerline, dimensioning to	3
Chamfer, sketch	1
Circles	1
Conics	9
Construction geometry	3
Convert entities	3, 9
Dimension, editing	1
Dimensions, naming	5, 12
Dimension, smart	1
Equations	12
Fillet, sketch	1
Intersection curve	3, 9
Lines	1
Mirror entities	9

Offset entities	3
Polygon	1
Rectangles	1
Scale	9
Silhouette entities	9
Sketch editing	1
Sketching on a plane or face	1
Sketch relations	1
Sketch tab	1, 3, 9
Slots	1
Spline	14
Tolerances	11
Trim entities	3

B.4) FEATURES

Command or concept	Chapter
Chamfer	1
Configuration, add	5
Cosmetic threads	11
Design table	5
Dimension names	5, 12
Dimensions, show feature	12
Dimension, show name	5, 12
Dome	9
Equations	12
Extrude boss/base, Extruded cut	1
Feature manager design tree	1
Feature suppression	12
Feature, editing	1
Feature tab	1, 3
Fillet	1
Hole wizard	3
If Statement	12
Lofted boss/base & Lofted cut	9
Mirror	3
Parametric modeling	12
Patterns	3
Reference geometry	3, 9
Revolve boss/base, Revolve cut	3
Rib	9
Rollback bar	1
Shell	9
Stud Wizard	11
Swept boss/base & Swept cut	9
Thread	11
Tolerances	11

B.5) VIEW

Command or concept	Chapter
Pan	1
Rotate	1
View cube	1
View (heads-up) toolbar	1
Viewing your part	1

B.6) MATERIAL AND PROPERTIES

Command or concept	Chapter
Material, applying	1
Properties, mass	1

B.7) DRAWING

Command or concept	Chapter
Annotation tab	2, 4
Assembly drawing	8
Ballooning	8
Bill of materials	8
Center line	2
Center mark	2
Datum feature	4
Detail view	4
Dimensions	2
Display style	2
Layers, Layer toolbar	2
Model items	2
New drawing	2
Note, leader	4
Projected view	2
Section view	4
Sheet format, edit	2
Sheet, edit	2
Sheet size	2
Sketch tab	4
Standard 3 view	2
View Layout tab	2, 4

B.8) ASSEMBLY

Command or concept	Chapter
Assembly visualization	10
Copy. part	7
Exploded view	7
Feature manager design tree	7
Fix	7
Float	7
Insert component	7, 13
Interference detection	10

Mate	7, 10
Mates, advanced	10
Mate, mechanical	13
Mates, standard	7
New assembly	7
New part	13
Sensors	10
Toolbox components	7

B.9) SIMULATION

Command or concept	Chapter
Add-Ins	6
FEA/FEM	6
New study	6
Report	6
Simulation study tree	6
Static simulation	6
Strain	6
Stress	6
Tensile strength	6
Yield strength	6

B.10) RENDERING

Command or concept	Chapter
Appearance	15
Camera	15
Final render	15
PhotoView 360	15
Preview window	15
Render region	15
Render tools tab	15
Rendering	15
Scene	15